CAMBRIDGE TEXTS
HISTORY OF PHILOS

GIORDANO BRUNO
Cause, Principle and Unity

CAMBRIDGE TEXTS IN THE HISTORY OF PHILOSOPHY

Series editors

KARL AMERIKS

Professor of Philosophy at the University of Notre Dame

DESMOND M. CLARKE

Professor of Philosophy at University College Cork

The main objective of Cambridge Texts in the History of Philosophy is to expand the range, variety and quality of texts in the history of philosophy which are available in English. The series includes texts by familiar names (such as Descartes and Kant) and also by less well-known authors. Wherever possible, texts are published in complete and unabridged form, and translations are specially commissioned for the series. Each volume contains a critical introduction together with a guide to further reading and any necessary glossaries and textual apparatus. The volumes are designed for student use at undergraduate and post-graduate level and will be of interest not only to students of philosophy, but also to a wider audience of readers in the history of science, the history of theology and the history of ideas.

For a list of titles published in the series, please see end of book.

GIORDANO BRUNO

Cause, Principle and Unity

TRANSLATED AND EDITED BY

ROBERT DE LUCCA

Duke University

Essays on Magic

TRANSLATED AND EDITED BY

RICHARD J. BLACKWELL

St Louis University

WITH AN INTRODUCTION BY

ALFONSO INGEGNO

University of Florence

CAMBRIDGE
UNIVERSITY PRESS

PUBLISHED BY THE PRESS SYNDICATE OF THE UNIVERSITY OF CAMBRIDGE
The Pitt Building, Trumpington Street, Cambridge CB2 1RP, United Kingdom

CAMBRIDGE UNIVERSITY PRESS
The Edinburgh Building, Cambridge CB2 2RU, United Kingdom http://www.cup.cam.ac.uk
40 West 20th Street, New York, NY 10011-4211, USA http://www.cup.org
10 Stamford Road, Oakleigh, Melbourne 3166, Australia

First published 1998

Typeset in Ehrhardt 11/13 [CP]

A catalogue record for this book is available from the British Library

Library of Congress cataloguing in publication data

Bruno, Giordano, 1548–1600.
Cause, principle, and unity/translated and edited by Robert de Lucca. Essays on magic/translated and edited by Richard J. Blackwell. A general account of bonding/[all three written by] Giordano Bruno; with an introduction by Alfonso Ingegno.
p. cm. – (Cambridge texts in the history of philosophy)
Includes bibliographical references and index.
ISBN 0 521 59359 X (hardback). ISBN 0 521 59658 0 (paperback)
1. Metaphysics – Early works to 1800. 2. Magic – Early works to 1800. I. Lucca, Robert de. II. Blackwell, Richard J., 1929– .
III. Bruno, Giordano, 1548–1600. Essays on magic. 1998.
IV. Bruno, Giordano, 1548–1600. General account of bonding. 1998.
V. Title. VI. Series.
B783.A4E5 1998
195–dc21 98–15503 CIP

ISBN 0 521 59359 X hardback
ISBN 0 521 59658 0 paperback

Transferred to digital printing 2003

Contents

Introduction

Giordano Bruno was born in Nola, near Naples, in 1548. He entered the Dominican Order and, following publication of some works that are now lost, he left Italy in 1579 for Switzerland, France and eventually England, a move perhaps due to the oppressive climate in his own country, where the church felt itself threatened by the new science which he attempted to propagate. Having acquired a great interest in Ramon Lull (*c.* 1232–1316)[1] and the art of memory, he presented in London his vision of an infinite universe in which he sought to re-unify terrestrial physics with celestial physics on the basis of a principle of universal becoming. He also reflected on the causes of the religious wars and tried to determine the origin of the theological disputes of the period. Beginning with the metaphysics expressed in *De la causa, principio e uno* (*Cause, Principle and Unity*), which reflected the objections he encountered in England, he derived a new concept of the divinity which evolved from his cosmology and was to assume a radically anti-Christian character. The magical, animistic vision of everything which he adopted throughout all his writings, not just those of the last period of his life, is evident here. In addition to his specific contributions to the scientific revolution, he presented a general metaphysical vision that contributed significantly to the development of Renaissance philosophy.

Having returned to Italy in 1591 during the debate about the legitimacy of combining ancient knowledge with orthodoxy, Bruno was perhaps deceived by the experience of Francesco Patrizi,[2] who was lecturing in

[1] Lull designed an *ars combinatoria*, a code for representing reality such that its elements could be combined in different ways to represent various items of knowledge, from astronomy to theology. Mastery of this code and its permutations provided the person trained in its use with a sophisticated mnemonic device.

[2] Francesco Patrizi (1529–97) was one of the leading Platonists of the Renaissance; his major work, *A New Philosophy of the Universes*, was condemned by the Congregation of the Index in Rome.

Platonic philosophy at the University of Sapienza at Rome. He thought he might be able to find a role for himself by renouncing or concealing the most heterodox features of his own teaching. This was an illusion, and he fell foul of the Inquisition and was executed at the stake in the Campo de' Fiori in 1600.

I

La Cena de le Ceneri (*The Ash Wednesday Supper*) was the first of the dialogues in Italian which Bruno published in 1584/5.[3] The striking feature of this work, in which the author proclaims his Copernicanism, is the immediate connection established between the annual motion of the earth around the sun and the infinity of the universe. This, however, was quite different from the position of Copernicus, who, having given new dimensions to the traditional cosmos, recognized the immensity of the heavens but left to the natural philosopher the ultimate decision about whether or not the universe was infinite. In *The Ash Wednesday Supper*, on the contrary, we find a clear affirmation of an infinite universe with infinite solar systems similar to our own. Suns and earths are composed of our own elements, they are living and inhabited beings, they are stars which are recognized not only as living things but also as divinities.

Bruno was led to these conclusions, in particular the thesis of the infinity of the universe, by a number of factors. In Copernicus' work, the earth was construed as a celestial body rotating round the sun like the other planets; it was implicitly elevated to the status of a star, thus breaking down the rigid separation between the sublunary world and the celestial world, although Copernicus did not want to confront the enormous physical problems which derived from his heliocentrism. It is significant that, in his *De revolutionibus orbium celestium* (*On the Revolutions of the Heavenly Bodies*), the sphere of fixed stars no longer had a specific physical function and no longer constituted the principle of motion. This was a conclusion that could have been strengthened in Bruno's eyes by some developments in Italian philosophy of nature, especially those of Bernardino Telesio (1509–88). Bruno now went further and called into question the very

[3] References to Bruno's Italian works are in the *Dialoghi italiani*, 3rd edn edited by G. Aquilecchia, reprinted with notes by G. Gentile (Florence: Sansoni, 1958: reprt. 1985). The Latin works, *Opera latine conscripta*, were edited in Naples between 1879 and 1891 in three volumes (in eight parts) by F. Fiorentino, F. Tocco, G. Vitelli, V. Imbriani and C. M. Tallarigo. References to the Latin works are identifed as *Op. lat.*, with the volume, part and page number.

existence of such a sphere, which seemed to him merely the result of an optical illusion which made all the stars appear to be at an equal distance from the earth.

Bruno's comparison between himself and Copernicus in *The Ash Wednesday Supper* throws further light on this issue. Although Copernicus is ranked in the history of astronomy as being comparable to Hipparchus or Ptolemy, his real significance is thought to lie in the fact that he is a hero of human thought who was able to oppose the force of common prejudice, the vulgar Aristotelian philosophy, the apparently self-evident view that the earth was immobile in the centre of the heavens. Nevertheless, his work is presented as having crucial limitations which open the way to what will be Bruno's specific contribution. Copernicus was primarily a mathematician – his interest was directed towards astronomy rather than towards natural philosophy, and in this sense his work needed to be further developed. Certainly he started from a correct and significant physical presupposition, the earth's motion, but he sought only a mathematical description of the movements of the heavens.[4]

In contrast, Bruno presents himself as a natural philosopher, as the one who is destined to become the authentic interpreter of Copernicus' discovery and is called to draw out the conclusions from it, beginning with the physical ones. The first of these, which is decisive for a correct understanding of the others, is the infinity of the universe. In the *Narratio* of Georg Joachim Rheticus, which Bruno was able to read in the 1566 edition of *De revolutionibus*, Rheticus had described the astronomer as a blind man who has a stick to help him on his way, and this stick was mathematics. In order to accomplish the theoretical task which he sets himself, a task which lies at the limit of human ability, the astronomer needs a hand to guide him and inspiration from above. Thus in *The Ash Wednesday Supper* Copernicus becomes the inspired one to whom the gods have entrusted a message, the importance and significance of which he has not realized; he is like a blind fortune teller for whom Bruno acts as the authentic interpreter. The philosopher, therefore, is summoned on a metaphorical journey across the heavens to discover that the traditional crystalline spheres are only a vain fiction, that there is no upper limit to the physical world and thus no end to his journey, and that what opens out in front of him is an infinite space. The philosopher shows us that the divinity is present in us and in our planet no less than in every other heavenly body, that it is not situated

[4] *Dialoghi*, 26–9.

beyond the imaginary limit of a closed and finite universe, in a place which makes it accessible to man.[5]

Bruno's reform, therefore, is not only philosophically significant but also has religious consequences. It challenges the developments of the Reformation, calls into question the truth-value of the whole of Christianity, and claims that Christ perpetrated a deceit on mankind. In the pages which follow, he compares the negative consequences which have resulted from traditional philosophy – negative consequences which are apparent to everyone – with the positive fruits, both civil and religious, which the new philosophy is producing, revitalizing all those fields of knowledge and life in which the ancients had excelled.

The consequences of this new philosophy are wide-ranging and radical because this new vision of the cosmos changes our relationship with the divinity, and this, in Bruno's eyes, transforms the very meaning of human life. He claims that this new vision will reconcile us with the divine law which governs nature, and free us from the fear of imaginary divinities, cruel and unfathomable, who look down from heavenly heights, controlling the sublunary world in a mysterious way. Human beings believe that they are enclosed in an inferior world subject to generation and corruption, but this is a simple illusion. Within this world, as in Plato's cave, we can see only the shadows of reality which appear on its wall, the shadows of the ideas which take shape and form at the upper limit of the heavens. Bruno suggests that, on the contrary, we can now recognize the universal law which controls the perpetual becoming of all things in an infinite universe. Knowledge of this law reassures us in the face of the present and the future (about which, of course, we have only an imperfect knowledge), because it does not deny anything its existence in and of itself, but claims that everything is being ceaselessly transformed into something else.

More than any previous thinker, then, Bruno is aware of the fact that the fall of Aristotelian cosmology implies the end of traditional metaphysics. From this starting point he elaborates a philosophy which is new and original, despite drawing on views attributed to the Presocratics (the *ens et unum* of Parmenides, Anaxagoras' *omnia in omnibus*), whose voices are distorted by the fact that they are preserved only in Aristotle's refutations of their positions. Thus, in *Cause, Principle and Unity*,[6] he sets about presenting a

[5] *Ibid.*, 32–4.

[6] See the critical edition of *De la causa, principio e uno*, edited by G. Aquilecchia (Turin: Einaudi, 1973).

metaphysics which is intended to constitute a more solid foundation for the interpretation of nature and for the consequent introduction of a new ethic, capable of establishing the outlines of the renewed relationship between man and God both at the level of civil life and at the philosopher's level of contemplation. The problem which immediately arises, however, is that of specifying how this new idea of the divinity is formed and in what sense Bruno's infinite universe radically modifies the relationship between God and the world, between God and human beings.

II

To clarify these issues, we must return to Bruno's earliest works, especially to *De umbris idearum* (*The Shadows of Ideas*) (1582). Here he tried to elaborate an art of memory which was based on magical foundations; and in doing this he identified the heavenly models, the exemplars of every sensible reality which the human mind can know, with the images of the thirty-six heavenly deacons which tradition attributed to Teucer the Babylonian and which he borrowed from the classic text of Renaissance magic, Agrippa's *De occulta philosophia*.[7] In *De umbris* Bruno applies, in an apparently arbitrary way, Nicholas of Cusa's coincidence of opposites to the conception of the hierarchy of being which Marsilio Ficino explained in his *Theologia platonica*.[8] This doctrine, which is central to that work, is an attempt to define the special privilege assigned within the framework of creation to the rational soul, a genus which includes both the *anima mundi* (the world-soul) and the human soul. Ficino defines this privilege in cosmological terms. In fact, in his eyes the rational soul was at the centre of the hierarchy of being, as the very link between the sensible world and the intelligible world; descending from the former, it gave life and form to the latter.

The hierarchy of being extended between two extremes, pure act and pure potency, God and prime matter, in such a way that each of the intermediate levels of the hierarchy presented a different relationship between act and potency. One descended down the levels of this hierarchy, starting

[7] Cf. E. Garin, 'Le «elezioni» e il problema dell'astrologia,' reprinted in Garin, *L'età nuova. Ricerche di storia della cultura dal* XII *al* XVI (Naples: Morano, 1969), 423–47, used, especially in ch. XI, by F. A. Yates, *Giordano Bruno and the Hermetic Tradition* (London: Routledge and Kegan Paul, 1964).

[8] M. Ficino, *Theologia platonica*, XI in *Opera* (Basel, 1576) I, 221–2. Ficino's doctrine is comprehensible due to the theory of the *primum in aliquo genere*, according to which the last member of one genus coincides with the first member of the following genus.

from the pure act constituted by God and eventually reaching prime matter. Each step downward represented an increase in potentiality. Within this overarching hierarchy, if the sensible and the intelligible are analysed as two separate categories and if each one of them is considered as a unitary whole, complete in itself, it would be possible to discover something new, namely the way in which the sensible world and the intelligible world, despite being radically distinct by nature, were linked together. In the intelligible sphere, one descended gradually to the lowest level, which was constituted by the rational soul; it was purely receptive to the levels above it, and could thus be considered as pure potency in relation to them. In the sensible sphere, on the other hand, one moved up within the hierarchy of being, from prime matter, through a sequence of more complex forms of corporeal organization until one reached an absolute limit. That limit was heavenly matter, which because of its purity and spirituality could be defined by Ficino as *corpus quasi non corpus* (a body that is almost not a body). This kind of matter, sometimes called 'spirit' and sometimes 'ether', could be considered to be pure act in comparison with prime matter. Here it seemed as if the pure potentiality which defined prime matter was transformed completely into its opposite, pure actuality. In conclusion, the more the act transformed itself into potency with respect to the superior levels in the intelligible world, the more the opposite process seemed to take place in the sensible sphere and potentiality seemed to be transformed progressively into actuality.

Here it is important to note how this analysis underpins Ficino's doctrine of the world-soul, which linked the corporeal and the spiritual, giving life and form to the entire inferior world. Bruno saw this as an instance of Nicholas of Cusa's coincidence of opposites: two spheres were gradually losing their essential characteristics by somehow transforming themselves into one another. He also saw in doctrines of this type the theoretical basis for a distinctive kind of art of memory and the foundation for an authentic astral theology. Through these it seemed possible that man, endowed with a rational soul and a spirit to mediate between the soul and his elementary body, could link himself to that privileged cosmic point on the boundary between the sensible and the intelligible which would allow him to grasp the archetypal forms, the actual generating models of every sensible reality, if not in their purity, then at least in their shadows, the shadows of ideas.

As already mentioned, in *The Ash Wednesday Supper* the sphere of fixed stars began to lose all the functions which had been assigned to it within traditional cosmology. Each of the movements which had been attributed to it was reduced to a mere appearance generated by the motion of the earth. Bruno thus denied the very existence of such a sphere, relegating it simply to an optical illusion. The first casualty of all this was Ficino's doctrine of the hierarchy of being, which Bruno had used in *De umbris*, where he interpreted it in terms of the coincidence of opposites; nevertheless, in this work he still tried to interpret the role of human beings, their origin and destiny, within the traditional cosmological framework. Certainly, he remained faithful even in his new cosmology to the Platonic world-soul, understanding it as an intrinsic principle of motion for all the celestial bodies which no longer needed any other forms of motion, and, as we shall see in *Cause*, he will speak of a universal soul which effectively shapes and gives life to everything. However, he is not able to refrain from attacking, in *De immenso* (*The Boundless*), those 'shadows of ideas' that men had believed in, all those *mysteria platonica et peripatetica* (Platonic and peripatetic mysteries) which resulted from the belief in two ontologically separated spheres, the heavenly world and the sublunary world. In particular, he summarizes and rejects all the characteristics attributed to the spheres of fixed stars which, among other things, made it the access route from the intelligible world to the sensible world.[9]

It is important, therefore, that he summarizes Ficino's doctrines of the hierarchy of being and of the meeting of the sensible and the intelligible in such minute detail in order to be able to reject them in a radical manner.[10] In the final, decisive book of the poem, he condemns both the theologian's empyrean heaven and the Platonic intelligible world, and undercuts the doctrine of spirit, conceived as an ethereal vehicle of the soul in its process of incarnation. The idea of a world of ideal moulds, of separated ideas, no longer has any meaning for him, and this rejection of a separate world of pure essences leads him to define as meaningless anything lacking a concrete, real existence, anything which, as a result of a process of abstraction, has been unjustifiably hypostatized.

Bruno's reflective transformation of Ficino's doctrine of the meeting between the sensible and the intelligible is essential for understanding the

[9] *Op. lat.*, I, II, 6: '... prima naturae genitura, simplicissima, capacissima, potentissima, activissima, animatissima, perfectissima, causa universalis ... cuius portae geminae ... divinarum animarum vehiculum, idearum characteribus signata ... nostro verenda metuendaque superincubans mundo, divinitatis potentia ... nunc spacii et aetheris natura, et magnitudine comperta ... e manibus, eque oculis evanescit, portentosa umbra sine corpore tandem fuisse convincitur.' For the reference to Macrobius, cf. *Op. lat.*, I, II, 150.

[10] *Op. lat.*, I, II, 116–18.

development of thought in *Cause*. If one starts from the assumption that the universe is infinite, it no longer makes sense to conceive the coincidence between act and potency as the exclusive property of a fixed point in the hierarchy of being, a privileged point in a finite and physical cosmos conceived as distinct from the intelligible world. Bruno therefore tries to rethink such a coincidence on the assumption that space is infinite and homogeneous, and that there are no separate hierarchical orders of being, and he does this in the light of two key concepts, that of an infinite active potency and that of an infinite passive potency, which are directly associated with each other in the cosmos. On this journey, Nicholas of Cusa guides him.

III

Nicholas of Cusa maintained, in *Docta ignorantia* (*Learned Ignorance*), that it was impossible to explain in conceptual terms the passage from the *complicatio* of everything in God to its *explicatio* in things; his recourse to the concept of 'contraction' to define the relationship between God and the universe has merely symbolic significance. It is not a real explanation, simply a suggestive way of referring to the inexplicable. The universe, *maximum contractum* (i.e. the limit of contraction), reproduces the unity of the divine in its proper form; it therefore is a coincidence between actuality and potentiality, although there is an insuperable limit to its actuality in the sense that the world can never realize its full potentiality. In fact, the only way the cosmos can realize its totality is through differentiation and spatial dispersion. The power to create and the power to be created coincide perfectly in the unity and absolute distinction of God; in contrast, the potentiality of the universe is a pale reflection of the infinite passive potency of God. And thus there will always remain an infinite difference between the 'contracted' existence of the universe and the unity and distinction which coincide in the divinity.

For Cusa, therefore, God and God alone was absolute possibility coinciding with absolute actuality. Despite its limits, the concept of contraction allowed him to conclude that the relationship between God and the world could never be explained by recourse to the philosophers' matter and the world-soul of the Platonists. Matter is possibility and if, as some have claimed, it is co-eternal with God, then it would become absolute possibility; it would then no longer be just something created by God, nor would it be contracted, as it in fact is, so as to give rise to a world of distinct entities. Bruno assigns to the Platonists' world-soul the role which it had in traditional cosmology, as mediator on the cosmic plane. This mediating role cannot be understood as the distinct possession of the exemplary models of all things, because this would imply that it displaced the Word, the only place in which the ideal archetypes rest in both absolute unity and absolute difference. Thus, the traditional ways of construing the world-soul and the relation between matter and the vivifying action of a universal spirit fail.[11]

Nicholas of Cusa outlines a cosmology which no longer recognizes ontologically separated levels in the universe. In the Cusan cosmos, everything is the centre and the circumference is nowhere – a distinction which Bruno considers a mere play on words. In this way, the earth loses the subordinate status which it had until now, in that it is thought to be no less central than any other star; it is subject to influences but is a probable source of influences itself. Cusa retains the traditional ontological inferiority of the heavens with respect to a divinity who holds them at an infinite distance from himself, and this is confirmed, in an apparent paradox, by the redemption of the earth. The fact that everything in the world is undergoing constant change implies that no absolutely precise relations exist and that we cannot have exact or real measures for any phenomenon, including motion.

This is the context for Cusa's Christology. If the distance between God and the universe is infinite, this can never be bridged by a mere man, even if he is exceptionally gifted; only the one perfect man, Christ, can achieve such a mediation through the Word, which leads creatures back to its source.

In *Cause*, Bruno drew the conclusion from his study of Cusa that nothing now prevents him from looking for the coincidence between the world-soul and the matter which belongs to an infinite universe as the coincidence of infinite active potency and infinite passive potency. Bruno conceives the hierarchy of being as having only ideal value, in contrast to Ficino's ontological conception of it, and he construes the world-soul and matter as the absolute opposites of this hierarchy. Starting from these assumptions, he tries to show how act and potency, absolute possibility and infinite actuality coincide. Thus it is only by starting with such a coincidence that he can apply the concept of 'contraction' to the relationship which is formed

[11] *N. Cusani De Docta ignorantia*, II, ch. VIII, IX, dedicated respectively to the possibility or matter of the universe, and the soul or form of the universe.

between the unity of the universe and the multiplicity in which this is structured.

Certainly, at the beginning of *Cause*, he warns that his discussion is meant to stay within the limits of pure natural reason, that it aspires to be only a philosophical discussion, leaving to theologians the more exalted task of defining the Prime Mover. But the route he follows is inevitably destined to hold some surprises in relation to such a cautious preliminary declaration. The coincidence between infinite active potency and infinite passive potency, which Nicholas of Cusa had recorded in *De possest* as a peculiarity exclusive to God, is transferred in *Cause* to the relation of absolute opposites in the cosmos, and knowledge of this coincidence gives us a proper understanding of the unity of substance.

IV

From this perspective, the logic which guides Bruno in *Cause* is clear. He conceives the intellect as a superior faculty of the world-soul that produces forms. This represents a significant lowering of the status of the intellect, albeit to the highest kind of faculty which can exist. The world-soul possesses intellect and does not therefore need a superior principle from which to draw forms. It should be added that it operates as an art which is intrinsic to matter, in contrast to human art which inevitably acts on the surface of matter already formed. The world-soul, therefore, shapes matter from inside because it possesses the actual models which allow it, as an authentic efficient cause, to be also a formal cause. Since it animates an infinite universe, and there is no part of the universe that is not animated or that does not possess at least a spiritual principle always capable of being actualized by it to some degree or other, differences in nature between the forms it gives are inevitably to be found.

The world-soul is therefore the authentic form of forms; it contains them all in act within matter and can therefore be considered either a cause or a principle, depending on whether we think of the forms as its possession or as superficial configurations that matter assumes now and again according to its dispositions. What is at issue here are the constantly changing forms of matter which the Aristotelians can only arbitrarily call forms in a strict sense. That is one of the constant features of the anti-Aristotelian polemic in *Cause*, because it becomes essential for Bruno to maintain that these are only appearances, which are constantly changing, compared with substance, which cannot be annihilated and is the active principle and producer of real, rather than transient, forms. This polemic against the supposed substantial forms of tradition is therefore already a vindication of the authentic active potency of an infinite universe, and opens the way to Bruno's special treatment of matter considered as potency. Then the confrontation with Nicholas of Cusa's theses becomes direct, although his name is never mentioned in this particular context.

Certainly, for Bruno, as for Cusa, it is only in God that infinite actualization of infinite possibility can be achieved. In the universe, on the other hand, things are constantly changing, and matter is inescapably subject to these changing forms. Despite this, the universe can be said to be completely infinite, to be all that it can be, provided one considers it as extended through all of time rather than at a single instant or from the point of view of eternity. However, the difference between God and the universe represents only the starting point of Bruno's discussion.

The power to be, if considered as passive potency, moves towards its infinite actualization only in God; in Him alone, act and potency, power to create and power to be created, are superimposed speculatively without reference to time and place. If, however, one considers matter absolutely as passive potency, if one abstracts it from the relationship which it has, at different times, with both corporeal and incorporeal substances, one notices a significant factor. There is no difference between the passive potency of these substances except for the fact that corporeal matter is contracted into dimensions, qualities, quantities, shapes, etc.; these accidental determinations (dimensions, shapes, etc.) are what the Peripatetic tradition, struggling to understand them, confused with genuine substantial forms. Dimensions, qualities, etc. do not, however, modify pure passive potency as such, and it is possible to conclude, therefore, that the matter which is conceived in these terms can be considered common to both the spiritual and the corporeal.

Bruno clinches his argument by referring to the Neo-Platonic doctrine that intelligible entities were composed of a very particular kind of intelligible matter. Such intelligible entities, which are forms of acting, must have something in common, although it cannot be anything that generates a distinction between them or involves any passage from potency to act. In the sensible world, where becoming involves such a passage, is not matter best understood as potency, which includes in its complexity all the dimensions and qualities, and does this not mean that this matter, rather than not possessing any form, in reality possesses them all? Could it be that matter, which appears not to produce distinctions, seems thus to be formless only because it is the origin of more deep-seated but less apparent distinctions – distinctions which it can be seen to possess only in a higher unity? Furthermore, this allows Bruno to claim that the two matters, the intelligible and the sensible, seen from the perspective of potency, can be reduced to a single genus, since the former is differentiated from act only by a distinction of reason and the latter can be considered act in comparison with the ephemeral and transient forms which appear and disappear on its surface. It would be impossible, then, to distinguish matter understood as potency from the world-soul.

Thus in this way Bruno assimilates his treatment of matter to the tradition of Aristotelianism and Neo-Platonism, which took matter to be a substrate, that which remains constant beneath the transformations which take place between the elements. In his eyes, the permanency of matter comes to mean that it, too, as the world-soul, is a principle which is neither passing nor transient, a principle which cannot be annihilated and which is identified with the substance of beings themselves. Bruno reminds us that the Aristotelians, as soon as they realized that they could not accept the Platonic solution which placed ideas outside the field of matter, admitted that matter could generate forms. Bruno called these ideas 'ideal moulds', and was more able to accept them than the Peripatetics were. It must be added that these same Aristotelians, when they state that matter passes from potency to act, speak only of the composite when specifying what has really changed. On the basis of all these elements, it seems legitimate to think that, if it is recognized as a constant and everlasting principle, prime matter cannot be classified as that *prope nihil* (almost nothing) of uncertain reality which figured in the views of a number of previous thinkers who tried to devise definitions of substantial form. These definitions, contrary to their intentions, all turn out to be reducible to pure logical abstractions. On the contrary, the fact that this matter presents no form would be equivalent once more, for the reasons already mentioned above, to its possessing all of them.

If, however, a spiritual principle and a material principle are recognized as the very substance of our world, it seems evident that it is their coincidence that constitutes its permanent substance. An analogous identification could then apply to the superior world of exclusively spiritual substance, which Bruno stated he would not discuss because he wished to confine his treatment to the limits of pure natural reason. This is the most ambiguous statement of the whole work, and understanding this ambiguity correctly is the key to understanding Bruno's philosophy. Bruno takes for granted here the separation which the whole dialogue tries to call into question, and at the most decisive point of the work, he refers to the notion of an intelligible matter of the superior world only to understand it in terms of corporeal substances seen from the perspective of potency. The ambiguity of such a statement allows him to leave an important fact in the background, that the relationship which he was establishing between infinite active potency and infinite passive potency created a relationship of reciprocal necessity between God and the world.[12] Thus Nicholas of Cusa's demonstration, in *De possest*, of the impossibility of separating, if only in God, the infinite potency of creating and the infinite potency of being created was decisive in forming Bruno's position. Bruno, however, came to the conclusion that these are present and inseparable in an infinite universe and that this involves not only their coincidence but, crucially, a relationship of reciprocal necessity between the unity to which they refer and the universe.

The solution rejected by Nicholas of Cusa and adopted by Bruno was, therefore, to return to the world-soul of the Platonists, and to a conception of matter as absolute possibility and as co-eternal with God, in order to explain the connection between all things in the cosmos. In fact, Bruno began from this conception of matter as absolute potency and from a world-soul which by now was the form of forms, and no longer required an ontologically superior principle to prepare exemplary models to inspire with its action. He thus discovered divine unity in their coincidence, a unity which preceded the distinction between the corporeal and the spiritual. This enabled him to set out the basic principles of his cosmology, which was different from Nicholas of Cusa's, but still based on the infinite distance, in terms of nature and dignity, between God and the universe. It thus became possible to imagine a mediation between the human and the divine which, moving through nature, would render unnecessary the solution adopted by Nicholas of Cusa and would in fact do away with all forms of Christology.

[12] He will begin to develop this point in *De l'infinito, universo e mondi*, concealing it slightly beneath the discussion of the relationship between God's *potentia absoluta* and *potentia ordinata*. *Dialoghi*, 385–7.

V

Bruno's originality lay in his rejection of that world of pure, ideal and bodiless essences. Arguing with the Platonists in the great conclusion of *De immenso*, Bruno states that there does not exist a justice separate from that which is good and, most importantly, that there is no divinity which can be distinguished from its manifestations.[13] Any attempt to make these distinctions is an unjustified hypostatization arising from processes of abstraction originating in our intellect. These are his final conclusions on the subject, which, when combined with the necessary nature of God's link to the world, constitute important keys to understanding *Cause*. If the universe is not contingent in its nature, it is possible to speak of a divinity which coincides with the world itself; this divinity would be a substance which from time to time manifests itself in infinite and different composites, in its 'modes', as Bruno calls them, which are themselves transient. Certainly, the unity to which multiplicity points as its foundation and its source remains in some sense absolute and not contracted, but the very fact that each part of the infinite is limited points to something which is the real condition of its existence. This means that one must conceive this unity as an internal unity of the cosmos rather than as something which is above or beyond it. The principle of the universe, if it is unique, is therefore its own cause, and this means that we cannot speak of two separate worlds. Thus, Bruno can state that God needs the world no less than the world needs Him,[14] since if the material infinity of the corporeal were lacking, the spiritual infinity of the divine would also be absent. By linking the world necessarily with the divinity and vice versa, the divinity is established as that which is all in all and in everything. It cannot be 'elsewhere', since its coincidence of spirituality with infinite matter means that 'elsewhere' does not exist.

Thus we arrive at the problem of understanding the unity of the All as an understanding of its laws in so far as they are laws of nature. Bruno is not mistaken here in claiming that the new departure he has initiated is radical. On the one hand, he believes he can demonstrate that both Aristotelian philosophy and the Christian religion, and not only the latter's most recent developments under the Reformation, have been linked to an erroneous cosmology. We need only consider the contemporary discussions on the ubiquity of the glorious body of Christ and the polemics concerning the

[13] *Op. lat.*, I, II, 310. [14] A Mercati, *Il Sommario del processo di G. Bruno* (Vatican City, 1942) 79.

nature of his presence in the Eucharist, both of which originated, according to Bruno, within the framework of this old erroneous cosmology. It is, therefore, understandable that this new philosophy should eventually reveal the full extent of its consequences and call for a healing of the division between nature and divinity decreed by Christianity; that it should search for laws, most notably in *Lo Spaccio de la bestia trionfonte* (*The Expulsion of the Triumphant Beast*), to constitute a new ethic, capable of guaranteeing peaceful civilian co-existence in the rediscovered harmony between human needs and the divine will. This same development of civilization can thus be reconceived according to those natural foundations which constitute its indispensable precondition. However, it is only by separating himself from these foundations, through a combined intellectual and physical effort, that man has been able to distance himself from the animal condition (symbolized in the myth of a terrestrial paradise) and bring himself gradually closer to God through science and the arts. It is not without significance that the fundamental error of Christianity, long before the Reformation, was the desire to begin with a divinity conceived in its absoluteness, arising from the illusion that in this way one could enter into contact with it and enjoy its favour, without respecting the intervening natural and cognitive levels. This general framework implies that Christ practised a deception when he promised men a transformation through which they could become sons of God, while in reality he was making them risk falling back into a purely animal condition by making the consumption of earthly food part of the sacrament of the Eucharist.

From this point of view, *Eroici Furori* (*The Heroic Frenzies*) acquires a particular importance, and also a religious one, in relation to the metaphysical theses of *Cause*. The contemplation of divinity which is realized in this work through the medium of nature is certainly destined by definition never to attain its final goal, the actual possession of the infinite. However, it is justified in that the 'enthusiast' encounters no upper limit to his contemplative ascent. Thus, *The Heroic Frenzies* concludes with one final philosophico-religious illumination: a vision of the kingdom of God and paradise, in which the human is transformed into the divine, in a metamorphosis to which not everyone can have access.[15]

The 'heroic enthusiast' comes to realize that he can translate everything into the species of his intellect, in a seemingly endless process of actualization. This is due to the bond of love which elevates him ever higher in this

[15] 'The *sursum corda*,' recalls Bruno polemically, 'is not in harmony with everyone.' *Dialoghi*, 1116.

process, eventually causing him to realize the infinite (and thus apparently illimitable) potentiality of his intellect. The process thus becomes an agonizing experience for the enthusiast because the more he retreats into himself, the more he is constrained by the magical force of love to come out of himself, to transform himself and live in the other, in a never-ending succession. In this way, the two opposites, act and potency, reveal not only their own coincidence but also the coincidence between intellect and love. Therefore, knowledge and love coincide with their object in the infinite; the intellect is transformed into the intelligible, the lover into the object of love. Knowledge and love are thus revealed as the two cosmic forces which are apparently separate in nature but which spring from the same potency and source.

VI

Given Bruno's earlier interest in magic and astrology, it is not surprising that the development of his new cosmology should introduce elements of uncertainty into his beliefs on these topics. In the notes left to us (which have been given the title *De magia mathematica*), he reconfirms, in a disagreement with Agrippa, his rejection of the traditional cosmic role attributed to the world-soul and to its ideas, and he rejects the physical action of stellar rays.[16] Whereas in *De immenso* he did not deny a symbolic value to the celestial bodies furthest away,[17] in *De rerum principiis* (*The Principles of Things*) he seems to reject even this value, at least for particular cases. In the same work he is critical of the astrological theory of aspects and of astrological books in general. He laments the confusion which has arisen due to the fallacious identification of planets with celestial bodies. He claims that the corruption which the magic arts have undergone with the passage of time has been due to the spread of error but also to a desire to keep the secrets of the arts out of the hands of the ignorant. Thus, he seems to be in favour of a reconstruction of planetary astrology which would have to take account of his new cosmology but which here appears to be only roughly mapped out. Within this tentative framework, which includes some elements of his new cosmology, he is still able to retain the astrological value of the traditional celestial images, apparently feeling that the observation of them continues to be useful and that they represent the survival of an ancient language.[18]

[16] *Op. lat.*, III, 503. [17] 'Multum valent signare, nihil causare remota.' *Op. lat.*, I, II, 265.
[18] *Ibid.*, III, 543–4.

All this throws light on some passages of *De magia*. Here Bruno, on the one hand, laments the extinction of that original and non-conventional hieroglyphic language in which signs designated things and apparently guaranteed communication with the divine; on the other hand, he preserves on the magical level the operational value of those characters, seals and figures which, according to tradition, propitiated demonic influence – it seemed possible not only to use them but also in some sense to remould them according to the dictates of a higher reason. More than once in his work Bruno tries to recreate something which elsewhere he claims has been irredeemably lost.

Bruno no longer accepts a separation between the natural, mathematical and divine worlds; therefore he can maintain a distinction between natural, mathematical and divine magic (or theurgy) only if he can posit the survival of a distinct object for each of these, without denying the possibility of a passage from one sphere to another. The stars have themselves become gods, in effect, and are inhabited by demons, while the divinity seems to occupy the infinite spaces which extend between worlds.

All this facilitates a process of interaction between natural and celestial magic, the most visible consequence of which seems to be the problematic nature of the distinction between the world-soul and the existence of a universal spirit. In other respects, the access to the divine world through the celestial seems to be linked to Bruno's natural philosophy and to the particular developments which his demonology had undergone.

Universal animism was what suggested to Bruno the schema according to which the whole of nature should operate and on the basis of which every type of magical operation should be modelled. Such a schema always provided for the action of an efficient universal principle, equipped with models of its action, on a passive principle. This holds true both in the action of elementary qualities, rendered perceptible to man and as a result of which one can legitimately speak of natural magic, and in the area of occult qualities ('occult' in the sense that they elude direct observation but are confirmed by the production of recurring causal links and of special effects which seem impossible to attribute to the action of elementary qualities). One has recourse in this case to the action of a universal spirit which was not necessarily located in the heavens of traditional magic. It is rather its particular corporeity which allows it to be extremely active and to produce all things, and Bruno clarifies the nature of its action by referring to the *corpora caeca* (blind bodies) which figure in Lucretius' *De rerum natura*.

The action of the magus at every level, therefore, consists in the preparation and modification of matter so as to render it susceptible to the desired influence. The world-soul has thus to be drawn into a portion of matter suitably prepared, so as to produce a particular effect. Precisely for this reason, the world-soul, which is present in all its entirety in everything, causes matter to be successively formed in an infinite variety of ways, and it does so according to specific principles of universal action. This property, of being totally present in everything, belongs also to several accidents of matter, like voice and sound, whose magic effect appears certain and whose action is ultimately attributable to the action of the soul. This allows one to explain several phenomena that were traditionally considered to be proofs of the existence of occult qualities, such as the attraction of iron by magnets, etc. Considering these phenomena, Bruno refers to a motion peculiar to matter which he terms 'spherical' and which consists in a body's acquisition or loss (*influxus* and *effluxus*) of minute particles of matter.[19]

Bruno uses the theory of a universal spirit not just to explain all recorded phenomena but also to delineate the specific features of his demonology. To him this spirit is the reason for the presence everywhere of living beings acting on us through means which elude the capacity of our senses. These can be subdivided into a number of species no less numerous than the number of living species on earth and differentiated from man by their superior or inferior faculties, as well as by their varying dispositions, favourable or not, towards us.

Since they act in a way which is imperceptible to our senses, it becomes essential to specify the point at which they gain purchase on our faculties, so that their influence can be avoided or repulsed. Bruno scornfully challenges the very successful *De occultis naturae miraculis* (*The Hidden Miracles of Nature*) of Levinus Lemnius,[20] and rejects a purely medical explanation of phenomena traditionally considered to be of demonic origin. His own explanation of such phenomena refers to both the inferior melancholic humour of the man who, because he is devoid of spirit, is especially vulnerable to demonic possession, and to the actual intervention of demons. These, possessing a body, affections and passions no less than man, are in search of whatever can constitute a source of nourishment or pleasure and, therefore, of a matter capable of attracting their action. What makes all of this possible is, on the one hand, the presence within us of a spirit which has a varying degree of purity, and, on the other, the fact that this spirit

[19] *Ibid.*, III, 418–19. [20] *Ibid.*, III, 45.

(whose link with our imagination can be taken for granted) is indistinguishable from the passive aspect of our consciousness. It is this faculty which may or may not allow the establishment of the demonic *vinculum*, depending on how much resistance the cognitive faculties are able to offer. According to the infinite diversity of physical constitutions and to the quality of the spirit which we can artificially (and sometimes wrongfully) modify, for example through certain foods or particular ointments, it is possible for a spirit to take control of us, attracted by our own melancholic humour, just as the world-soul can be attracted by a matter which is disposed to receive a certain influence. The demon thus becomes the cause of our deception, making appear as real what are simply ghosts of our imagination and even giving us the illusion of entering into contact with divinities who are also imaginary. On this basis, in *On Magic* and *Theses on Magic*, Bruno posits two types of humanity, one superior and one inferior to the general level of mankind, who are distinguished by their ability (or lack thereof) to monitor and direct the processes of our consciousness and in particular its inevitably passive aspect. This, of course, is one of the constant themes of his philosophy and in particular of his polemic against the Reformation. In addition, it illustrates his belief that real processes and cognitive processes have a common foundation which has a magical aspect. Since the publication of *Sigillus sigillorum* (*The Figure of Figures*), he had been proclaiming, in overtly religious terminology, the essential value of a *regulata fides* (regulated faith), that is, the importance of exercising conscious control over our receptive faculties. In this way, he argues against those '*qui aguntur potius quam agant*' (who are acted on rather than act).[21]

Bruno distinguishes between two types of contraction achievable by man. Contraction is a phenomenon through which the soul, by concentrating on itself, can realize particular powers; but this can have an opposite effect if it is directed towards a higher contemplative level or if it is carried out so as to render us no longer masters but servants of our imagination, and thus exposed to demonic influence. Here Bruno echoes Ficino in his exemplification of various types of contraction; but instead of calling them '*vacationes animi*', as Ficino had done, he gives them a name which allows him to incorporate this phenomenon into the metaphysical structure governing our consciousness.[22]

[21] *Op. lat.*, II, III, 193.

[22] M. Ficino, *Theologia Platonica*, XIII, 2, in *Opera*, (Basel, 1576) I, 292–5. Cf. *Op. lat.*, II, II, 180–93; the distinction between two opposite types of *contractio* is connected to the distinction between two types of melancholy. Cf. on this point R. Klibansky, E. Panofsky, F. Saxl, *Saturn and Melancholy* (London: Nelson, 1964).

The point of distinction between the two forms of contraction is therefore represented by the intermediate cognitive faculties which turn the data of sensibility into figments of our imagination. This distinction, and the separation into two distinct levels of humanity, find their exemplary expression in the *Cabala del cavallo pegaseo* (*The Cabala of Pegasus*) and in *The Heroic Frenzies*. The *Cabala* outlines the characteristics of the man who, faced with the difficulty of searching for the divine, freely renounces his superior faculties, those which make us really human, and contracts his cognitive powers into the single one of hearing, to passive reception alone. Thus stripped of all power of judgment and reduced to the animal condition of an ass, he can no longer tell if his rider is a god or a demon – an allusion to a famous line from Luther's *De servo arbitrio*, aimed at denying the very possibility of our freedom. This is the reason why, in *The Heroic Frenzies*, he praises the 'divine seal' of the 'good contraction'.[23] We have seen that, in this work,[24] the metaphysics of *Cause* are translated in terms of the highest experience which man can have, of contemplation of the divine by means of an adequate image of it. Bruno claims, however, that this can be attained only by someone whose mind is constrained by two bonds (*vincula*): love, and the highest intelligible species which divinity could present to his eyes (i.e. beauty and the goodness of nature). In relation to the action of these two *vincula*, the 'divine seal' of the 'good contraction' acquires an essential importance: divinity, in fact, yields and communicates itself to us only at a level proportionate to our receptivity of it. Therefore, it is always our responsibility to intervene in the passive moment of our consciousness so as to raise ourselves above that moment, actualizing the infinite potency which is within us.

This leads Bruno back to the distinction between two types of humanity, those who fall victim to demonic deception and those who, rising above the level of the multitude, overturn the scale of values in which humanity believes and set out to attain the level of a heroic humanity. A fascination with the Epicurean ethic which was already present in *The Heroic Frenzies*[25] appears here, in the works on magic, although this is a sophisticated Epicureanism that emphasizes the superiority of the learned man over every event. This man attains a different kind of mind – in fact, a different kind of spirit – and goes to meet a different destiny, while for the others, those who descend below the level of the mass of humanity, the

[23] *Dialoghi*, 877–9. [24] *Dialoghi*, 797. But cf. *ibid.* 1091–2.
[25] *Op. lat.*, III, 657. Cf. *Dialoghi*, 1052–54.

servitude of their own imagination can become a real hell on earth and can be indefinitely prolonged through reincarnation.[26] With *De vinculis in genere* (*A General Account of Bonding*), however, we seem to encounter a different picture of the fundamental problems discussed so far. The magus is acquainted with the dynamics not only of magic but also of demonic action, and knows how demons can take possession of us through unguarded avenues, and this opens up to him a new field of action, permitting him to link other men to himself and, in fact, to establish a whole series of magical bonds between himself and others. The moral problem raised by magic in general seems to take on a new aspect here. At the beginning of *On Magic*, Bruno examines the stereotypical moral objections which are advanced against magic in general, and against 'mathematical' magic in particular. His reply is equally traditional: magic understood as pure knowledge, as *scientia*, is always positive but it can be used well or badly, for good or evil, depending on who sets it to work. All this could be equally applied to *Bonding*; however, there seems to be a new element here which may raise a question, if not about the nature of Bruno's philosophy, then certainly about several of its characteristic features. This is a philosophy aimed at liberating man from the fear of death and of the gods, pointing the way to an escape from the snares which demons use to catch us. And yet here we find talk of the establishment of occult snares designed to put one man in the power of another, making the latter a kind of demon with the power to take possession of the other's spirit. It should be added that none of the effects attainable by man seems to be excluded from the scope of an action which, far from limiting itself to mere rhetoric, is meant to infiltrate every sphere of civil life. Certainly, Bruno's terminology continues to be traditionally magical; even Campanella was later to write a *Bonding* of his own in *De sensu rerum* (*On Sensation in Things*). It should be added that Bruno was an heir, albeit in his own original way, to one of the most important (and most fruitful) aspects of Italian speculation in the 1500s, namely the unprejudiced and often brutal observation of reality that is to be found in writings from Machiavelli to Cardano. There is still a tension here between Bruno's radically aristocratic vision and the fact that his work deals with what he believes are laws of nature, which provide no barriers in principle to universal ascent.

Bruno claims that the *vinculum* in itself is neither good nor evil, but the

[26] Cf. on this topic R. Klein, *L'enfer de Ficin*, *La forme et l'intelligible. Ecrits sur la Renaissance et l'art moderne* (Paris: Gallimard, 1970), 89–124.

fact remains that it presupposes a will to act on the part of the agent and a predisposition in the consciousness of the other person to be acted on in an occult and imperceptible way. All the bonds, he tells us, can be reduced to the bond of love, and this gives rise to a series of extremely acute observations which primarily affect the idea of beauty as conceived by the Platonists. They are observations which appear also to reveal a sort of intolerance towards a philosophical tradition which divided nature into diverse faculties, in particular the tradition which divided human nature into intellect and will. The *vinculum*, he says, is not found in the visible species, but what renders it active and often detrimental to us is something of which we are not aware, although it is sentient and active within us. It is precisely the difficulty of defining a single essence of love, of beauty and of pleasure which indicates to us that there are many different ways in which we can link with (*vincere*) the soul of the other. In order to put this binding process into action, we require a knowledge of the infinite variety of subjective and objective factors (beginning with the diversity of physical constitutions) in relation to which the *vinculum* must be prepared in advance in order to be effective. These elements, however, given that they exist in infinitely varied individual configurations, cannot be reliably specified in any given case. In this, they recall some of the central theses of Brunian metaphysics.

When Bruno outlines in *De immenso* the contemplation worthy of the perfect human being,[27] he takes inspiration from the image which he has of the divinity. The divinity is a matter which creates all and becomes all; thus, the perfect human being is one who, by elevating himself to the infinite in contemplation of the divine, actualizing in the infinite his cognitive potency, is capable of assimilating everything because he knows how to transform himself into it. The excellence of this *magnum miraculum* which is man is not taken for granted at the outset but rather constitutes a point of arrival and a final achievement. It coincides with the process of human deification, made possible by man's capacity to become, in some sense, *omniformis*, like divinity. It is therefore significant that, in *Bonding*, the metaphysical conclusions of *Cause* are taken up – the identity of *facere* and *fieri*, of the potency of creating and being created.

This metaphysical view not only implies that there exists no spiritual world which is separated from its corporeal support, but also implies that reality is unique, and this has important consequences for the psychological possibility of magical action.[28] This general scheme provides for two

[27] *Op. lat.*, I, I, 205–6. [28] *Op. lat.*, III, 695–6. Cf. *Dialoghi*, 262 and 315.

constituent moments, one active and one passive, where the latter has to be modified in order to make the former operational. Now, the mid-point between these two moments is, in fact, the *vinculum*, that which links to an ever-changing degree the operator (the *vinciens*) to the *vinciendum*. The original unity of the All, therefore, establishes the conditions for the success of magical action, because it allows us to understand how a magus can restore an existing apparent multiplicity to its underlying unity. Human beings, too, are presented as matter over whose surface pass infinite forms, and clearly each one of them is a *vinculum*, one of the many which we all, in fact, encounter. If we can give the right form to things we encounter, we can begin to operate on them according to the same magical scheme which we have found to be in operation on every other level of nature. This process can be guided artificially but does not go beyond the framework of nature, since it does no more than encapsulate in a unique form what are the guiding laws of nature itself. Once again, this is the myth of metamorphosis, that metamorphosis of all things which made possible on the operational level the recognition of the unity which underlies all things and their development. The action which one exercises on oneself (thus making oneself somehow one's own object) is aimed at transforming oneself into a subject of an ever higher form. Magical action is another instance of the coincidence between act and potency which the supreme contemplator has translated into the ability to become *omniformis* and which here, because of the potency of the *vincula* and, in particular, the most powerful of them all (love), is the ability to transform the other by actualizing the potency which is within him. One's action will thus have various levels according to one's capacity to give form to that potency by which one is linked to the *vinculum*. Finally, at the highest level, the *vinculum* reveals its deepest nature, transforming potency into act, act into potency, whence it follows that the operator is transformed in his turn into an object, and the *vinciendum* into *vinciens*.

Chronology

1548	Born at Nola, near Naples
1572	Ordained priest in the Order of Preachers (Dominicans). Began studies in theology
1576	Fled to Rome following proceedings brought against him for serious dissent about dogmatic theology
1579	Following several stays in northern Italian cities, went to Geneva where he became a Calvinist. However, he was charged with defamation and threatened with excommunication. He admitted his guilt and was pardoned
1581	Having taught at Toulouse, went to Paris. Interested the French court in his theory of memory and maintained contact with the court for five years, due to close links with the *politiques* who supported the King of Navarre. *De Umbris Idearum* (*The Shadows of Ideas*) (1582), which was dedicated to Henry III, *Cantus Circaeus* (*The Circean Melody*) and the Italian play, *Candelaio* (*The Candle Maker*), were published during this period
1583	In England as guest of the French Ambassador to Elizabeth I, Michel de Castelnau, perhaps entrusted with a political mission. Proposed Copernicanism in public lectures in Oxford, and introduced the philosophical and scientific themes of subsequent works in Italian. Rejected by the academic circles at Oxford, he returned to London where *Sigillus Sigillorum* (*The Figure of Figures*) was published
1584	In London, at the house of Fulke Greville, expounded the Copernican theory in a debate which is echoed in the first of

his Italian dialogues, *La Cena de le Ceneri* (*The Ash Wednesday Supper*). The debate provoked opposition, but did not damage his relations with Philip Sidney and the circle of Robert Dudley, Earl of Leicester. Bruno later defends himself in the first dialogue of *De la Causa, principio e uno* [*Cause, Principle and Unity*]

1584–5 Published, in London, the Italian dialogues: *La Cena de le Ceneri*; *De la causa, principio e uno*; *De l'infinito, universo e mondi* (*The Infinite, the Universe, and Worlds*); *Lo Spaccio de la bestia trionfante* (*The Expulsion of the Triumphant Beast*); *Cabala del Cavallo Pegaseo* (*The Cabala of Pegasus*); *Eroici furori* (*The Heroic Frenzies*) – all published by J. Charlewood with an incorrect place of publication. *Expulsion* and *The Heroic Frenzies* were dedicated to Sir Philip Sidney

1585 Returned to Paris, where he found a changed atmosphere which was unfavourable to him. Disputed the one hundred and twenty *Articuli de natura et mundo adversus peripateticos* (*Articles about nature and the world against the Peripatetics*) at the College of Cambrai; these articles were rewritten and published at Wittenberg under the title *Camoeracensis Acrotismus* (1588)

1586 At Wittenberg, where he gave lectures on the *Organon*

1587 Published a series of Lullian works

1588 Went to Prague, then to Helmstedt, where he remained until April 1590, despite disputes with the Lutherans and a new excommunication. *De Rerum Principiis* (*On the Principles of Things*) was sketched or finished during this period, and the works on magic, *De Magia; Theses de magia, De magia mathematica* (*On Magic*; *Theses on Magic*; *Mathematical Magic*), were completed, together with *De Vinculis in genere* (*A General Account of Bonding*)

1590 Went to Frankfurt to await publication of the three great Latin poems, *De Minimo*; *De Monade*; *De Immenso* (*On the Minimal; On Monads; On the Boundless*) (Wechel, 1591)

1591 During a second stay at Frankfurt, received an invitation from the Venetian patrician, Giovanni Mocenigo, to go to Venice to teach him the secrets of his art of memory. In Venice during August, perhaps hoping to get the chair of mathematics left vacant since 1588 (to which Galileo was subsequently appointed). A climate of hope for toleration prevailed in Europe, and perhaps the teaching of Francesco Patrizi at La Sapienza, Rome, deluded him about the possibility of enjoying a reprieve in Italy

1592 Imprisoned following three denunciations by Mocenigo to the Holy Office. The Venetian phase of his trial, which is well documented, was thus initiated; Bruno defended himself, claiming that his teaching was purely philosophical, that he was penitent and was prepared to renounce his errors

1593 Confined in the Roman jail of the Holy Office; the Roman Inquisition had obtained, with some difficulty, a transfer of the trial from the Venetian Senate

1594 Following a new denunciation and new depositions, Bruno's position became acute. He re-affirmed the line of defence adopted in Venice and presented a lengthy submission of eighty pages (since lost) which was a turning-point in the trial towards an unfavourable outcome

1596 A commission of theologians examined his published works which had not previously been used, to censure heretical propositions which they allegedly included and to report them to the trial. Included were propositions concerning the first principles of reality, the necessary connection between an infinite cause and an infinite effect, the conception of the individual soul and its relationship with the world-soul, the motion and soul of the earth, the identification of angels with the stars and of the Holy Spirit with the world-soul, and belief in pre-adamites

1598 Summary of the trial ready

1599 After a long interruption, trial re-activated; on the suggestion of Cardinal Bellarmine, eight heretical propositions were submitted to him for his unconditional repudiation. In a series of petitions and depositions, he claimed that he was agreeable to the renunciation; however, he also became entangled in the merits of the incriminating propositions by making various distinctions. Thus, his position deteriorated until the tribunal required him to acknowledge his errors. On

21 December, he said he would not agree to retract and that he did not know what should be retracted

1600 On 20 January, Clement VII ordered that he be condemned as an 'impenitent, stubborn and obstinate' heretic. The sentence was read to him on 8 February; it listed among his errors the denial of transubstantiation, the thesis of the transmigration of souls, the infinity of the world, the eternity of the universe, the allegation that Moses and Christ were magicians and impostors, and belief in pre-adamites. On 17 February, he was burned alive in Rome at the Campo de' Fiori

Further reading

The Latin works of Bruno are found in *Opera latine conscripta*, 3 vols in 8 parts, ed. by F. Fiorentino *et al.* (Naples: Morano, 1879-91), reprinted by Frommann, Stuttgart-bad Cannstatt, 1952. The Italian works are collected in *Dialoghi italiani. Dialoghi metafisici e dialoghi morali nuovamente ristampati con note da G. Gentile*, 3rd ed. edited by G. Aquilecchia (Rome and Florence: Sansoni, 1958). Other works by Bruno are *Candelaio*, ed. by V. Spampanato (Bari: Laterza, 1923); *Due dialoghi sconosciuti e due dialoghi noti*, ed. by G. Aquilecchia (Rome: Ediz. di Storia e Letteratura, 1957); *Praelectiones geometricae e Ars deformationum*, (Rome: Ediz. di Storia e Letteratura, 1964). G. Aquilecchia has also provided a critical edition of *La Cena de le Ceneri* (Turin: Einaudi, 1955), and of *De la Causa, principio e uno* (Turin: Einaudi, 1973). There is an Italian translation of the Latin poems by C. Monti, *Opere latine* (Turin: UTET, 1980).

There are bibliographies by V. Salvestrini, *Bibliografia di G. Bruno, 1582–1950*, edited by L. Firpo (Florence: Sansoni, 1958), and R. Sturlese, *Bibliografia censimento e storia delle antiche stampe di G. Bruno* (Florence: Leo S. Olschki, 1987).

Among the works of Bruno in English translation are *The Expulsion of the Triumphant Beast*, trans. A. Imerti (New Jersey: Rutgers University Press, 1964; rprt. Nebraska: University of Nebraska Press, 1994); *The Ash Wednesday Supper*, trans. S. Jaki (The Hague: Mouton, 1975); there is another edition of the same work, trans. E. Gosselin and L. Lerner (Hamden, Conn.: Archon Books, 1977; rprt. Toronto: Toronto University Press, 1995); *Cause, Principle, and Unity*, trans. J. Lindsay (Westport, Conn.: Greenwood Press, 1976); *On the Composition of Images, Signs and Ideas*, trans. D. Higgins (New York: Willis Locker and Owens, 1991); *The*

Heroic Frenzies, trans. P. Memmo (Chapel Hill: University of North Carolina Press, 1964; rprt. 1981).

Works about Bruno in English include F. Yates, *Giordano Bruno and the Hermetic Tradition* (Chicago: University of Chicago Press, 1990); F. Yates, *The Art of Memory* (London: Routledge and Kegan Paul, 1966); D. Singer, *Giordano Bruno: His Life and Thought: With Annotated Translation of his Work, On Infinite Universe and Worlds* (New York: Greenwood Press, 1968); P. Michel, *The Cosmology of Giordano Bruno*, trans. R. Maddison (London: Methuen, 1973); William Boulting, *Giordano Bruno: His Life, Thought and Martyrdom* (New Hampshire: Ayer Company Publishers, 1977); J. Bossy, *Giordano Bruno and the Embassy Affair* (New Haven, Conn.: Yale University Press, 1992); N. Ordine, *Giordano Bruno and the Philosophy of the Ass* (New Haven, Conn.: Yale University Press, 1996); A. Koyré, *From the Closed World to the Infinite Universe* (Baltimore: Johns Hopkins University Press, 1968); B. Copenhaver and C. Schmitt, *A History of Western Philosophy 3. Renaissance Philosophy* (Oxford: Oxford University Press, 1992); C. Schmitt, Q. Skinner, E. Kessler, and J. Kraye, *The Cambridge History of Renaissance Philosophy* (Cambridge University Press, 1988).

Other useful works about Bruno include F. Tocco, *Le opere latine di G. Bruno esposte e confrontate con le italiane* (Florence, 1889); G. Gentile, *G. Bruno e il pensiero italiano del Rinascimento* (Florence: Sansoni, 1925); V. Spampanato, *Documenti della vita di G. Bruno* (Florence, 1933); A Mercati, *Il Sommario del processo di G. Bruno* (Vatican City, 1942); A. Corsano, *Il pensiero di G. Bruno nel suo svolgimento storico* (Florence: Sansoni, 1940); L. Firpo, *Il processo di G. Bruno* (Naples: Ed. Scientifiche Italiane, 1949); N. Badaloni, *La filosofia di G. Bruno* (Florence: Parenti, 1955); H. Védrine, *La conception de la nature chez G. Bruno* (Paris: Vrin, 1967); F. Papi, *Antropologia e civiltà nel pensiero di G. Bruno* (Florence: La Nuova Italia, 1968); E. Garin, 'Le «elezioni» e il problema dell'astrologia' in Garin, *L'età nuova. Ricerche di storia della cultura dal* XII *al* XVI *secolo* (Naples: Morano, 1969); R. Klein, *La forme et l'intelligible. Ecrits sur la Renaissance et l'art moderne* (Paris: Gallimard, 1970); G. Aquilecchia, *G. Bruno* (Rome: Istituto della Enciclopedia Italiana, 1971); P. R. Blum, *Aristoteles bei G. Bruno* (Berlin: W. Fink, 1980); P. Rossi, *Clavis universalis: Arti mnemoniche e logica combinatoria da Lullo a Leibniz* (Bologna: Il Mulino, 1983); G. Aquilecchia, *Le opere italiane di G. Bruno* (Naples: Bibliopolis, 1993).

Note on the texts

Cause, Principle and Unity (*De la causa, principio e uno*) was first published in 1584 in London, during Bruno's sojourn there (1583–5). Few copies of the original printing survived and no other editions of the work are listed until the nineteenth century, when two important editions of Bruno's works were published by Adolfo Wagner, Lipsia, in 1830, and by Paolo Lagarde, Gottinga, in 1888. Thereafter, the book was frequently reprinted, either in whole or in part, both in Italian and in various translations, most notably as part of the critical edition of Bruno's works edited by G. Aquilecchia, *Dialoghi italiani* (Rome and Florence: Sansoni, 1958). The present translation is based on the text published in *Opere di Giordano Bruno e Tommaso Campanella*, edited by A. Guzzo and R. Amerio (Milan and Naples: Ricciard, 1956).

The translations of the *De magia* and of the *De vinculis in genere* are based on the texts published in *Jordani Bruni Nolani opera latine conscripta publicis sumptibus edita*, edited by F. Tocco and E. Vitelli (Naples and Florence: Morano, 1879–91), Vol. III, pp. 395–454 and 653–700 respectively (a shorter, earlier version of the *De vinculis* is found on pp. 637–52.) The Tocco-Vitelli edition was based on the text of the Noroff codex in Moscow, which was transcribed by Bruno's disciple Girolamo Besler, or Bisler, of Nuremberg between 1589 and 1591. Albano Biondi's Latin–Italian edition, *De magia, De vinculis in genere* (Pordenone: Edizioni Biblioteca dell'Immagine, 1986) was very helpful, and was consulted throughout the preparation of these first English translations.

Cause, Principle and Unity

Prefatory Epistle

Addressed to the most illustrious
Monsieur Michel de Castelnau
Seigneur of Mauvissière, Concressault, and Joinville,
Chevalier of the Order of the most Christian King,
Counsellor of his Privy Council,
Captain of fifty men at arms
and Ambassador to the most serene Queen of England.

Most illustrious and honoured Chevalier, if I consider with an appreciative eye the forbearance, perseverance and solicitude with which, adding favour on favour, benefit on benefit, you have bound, obliged, and tied me to you, and with which you are wont to prevail over every hardship, elude all sort of peril, and successfully conclude all your most worthy designs, I cannot but note how very appropriate is that noble device which adorns your terrible crest. On it a liquid humour sweetly strikes, with its constant and continual drip, and, by force of perseverance, softens, hollows, breaks, smooths and conquers a firm, solid, rugged and harsh rock.[1]

If, moreover (passing over all your other noble accomplishments), I recall how much you are for me, by divine commandment, by high providence and predestination, a firm and able defender against the unjust injuries that I suffer (and that wanted from me a truly heroic spirit in order not to throw up my hands, surrender to despair and succumb before the swift flood of criminal falsehood with which I have been furiously attacked, by the envy of the ignorant, the presumption of sophists, the deprecation of the malicious, the badmouthing of varlets, the insinuations of mercenaries,

[1] The device is the adage 'Gutta cavat lapidem'.

gainsaying of servants, suspicion of fools, slanderers' gossip, hypocrites' zeal, barbarians' hatred, the fury of the mob, frenzy of the populace, the complaints of those I have grazed and the cries of those I have scourged – in which there was not lacking the mean, frenzied and spiteful disdain of a woman, the false tears of whom are frequently more powerful than the stoutest waves and rudest tempests of presumption, envy, deprecation, slander, insinuation, betrayal, outrage, disdain, hate and fury), you appear to me then like a solid, secure and immovable reef which, rising up to show its crest above the swollen sea, is neither eroded, nor rent, nor moved by the seething heavens, nor by the dread of winter, nor by the violent crash of thick waves, nor by the harsh gusts of wind, nor by the wild blowing of the north wind, but rather is increasingly covered with greenery which clothes and adorns its flanks. You who are then endowed with that double virtue, which renders so mighty the mild and liquid drops, and so futile the blustery and rough waves, you through whom the lordly rock is so weakened beneath the rain and the tormented reef rises so powerfully against the flood, you are the one who offers both a secure and calm haven for the true Muses, and a deadly shoal on which the false ammunition and impetuous designs of the enemy sails are shattered. I, then, whom no one has ever succeeded in accusing of ingratitude or taxing with discourtesy, I, against whom no one may rightly complain, I, hated by fools, slighted by the contemptible, profaned by knaves, vituperated by rogues, and persecuted by brutish spirits, I who am loved by the wise, admired by the learned, glorified by the great, cherished by the mighty and favoured by the gods, I who have already gained such indulgence from you as to be received, nourished, defended, freed, placed in surety, sheltered at port, as of one who, thanks to you, has fled a great and dangerous storm, it is to you that I consecrate this anchor, these shrouds, these battered sails, these goods, to me most dear, and to future generations most precious, so that, thanks to your beneficence, they may not be submerged by the iniquitous and tumultuous Ocean which is my foe. Hung in the sacred temple of glory, by their power against the effrontery of ignorance and the voracity of time, they shall render eternal testimony to your invincible magnanimity; so that the world may know that, thanks to you, this bountiful and divine progeny, inspired by lofty intelligence, conceived by a tempered spirit and born of the Nolan Muse, has not passed away in its infancy, and will live as long as the earth, whose surface is so full of life, turns beneath the eternal regard of the other shining stars.

Here, then, is that sort of philosophy where one discovers, with truth and confidence, that for which we look in vain in diverse or opposing philosophies. First, then, I offer you a summary of five dialogues, which contains all that seems relevant to the effective contemplation of *Cause, principle, and unity*.

Argument of the first dialogue

In the first dialogue, you have something that you may call an apology, or what you will, concerning the five dialogues that make up *The Ash Wednesday Supper*, etc.[2]

Argument of the second dialogue

In the second dialogue you have, *first*, the cause of the difficulty of such knowledge, in order to know how far removed the knowable object is from the cognitive power. *Second*, in what manner and to what extent the cause and the principle may be explained by the thing caused or principled. *Third*, what the knowledge of the substance of the universe contributes to the conception of that on which the substance depends. *Fourth*, by what specific means we try to know the first principle. *Fifth*, the difference and accord, the identity and diversity existing between the meanings of the terms 'cause' and 'principle'. *Sixth*, the nature of that cause which we divide into efficient, formal and final; the different ways of defining the efficient cause, and from how many points of view it may be conceived. How this efficient cause is, in a sense, intrinsic to natural things, since it is nature itself; and how, in a sense, it is extrinsic to them; how the formal cause is joined to the efficient cause, and is that through which the efficient cause operates, and how the formal cause, itself, is brought forth from the womb of matter by the efficient cause; how the efficient and formal causes coincide in an elementary substratum, and how the one cause is distinct from the other. *Seventh*, the difference between, on one hand, the universal formal cause, which is a soul through which the infinite universe (insofar as it is infinite) is animated, not positively but negatively, and, on the other hand, the particular formal cause, multipliable and multiplied to infinity, which is more perfect insofar as it is found in a more general and superior substratum,

[2] Bruno's celebrated dialogue *The Ash Wednesday Supper*, written, like *Cause, Principle and Unity*, during his Oxford sojourn, 1583–5.

so that the great animals such as the stars must be fully considered as being more divine, that is endowed with an infallible intelligence and an activity without defect. *Eighth,* that the first and principal natural form, the formal principle and efficient nature, is the soul of the universe, which is a vital, vegetative and sensitive principle in all things which live, vegetate and feel. And by way of conclusion, that it is, moreover, unworthy of a rational subject to believe that the universe and its principal bodies are inanimate, seeing that from the parturitions and excrements of those bodies derive the animals that we call most perfect. *Ninth,* that there is nothing so defective, unfinished, abortive and imperfect that, since it has a formal principle, it does not likewise have a soul, even if it does not possess the act of substance which we describe as animal. And we may demonstrate, with Pythagoras and others who have not opened their eyes in vain, how an immense spirit, under different relations and according to different degrees, fills and contains the whole. *Tenth,* it is shown that, since this spirit exists unalterably together with matter (called 'shadow' by the Babylonians and Persians), and since both are indissoluble, it is impossible that, in terms of substance, anything can know corruption, or finish by dying; although, in terms of particular accidents, everything changes aspect and is transformed into now one composition, now another, abandoning and then taking up again now this being, now that. *Eleventh,* that the Aristotelians, the Platonists, and other sophists have not recognized the substance of things; and it is clearly shown that in natural things, all that they call substance, apart from matter, is nothing but the purist accident. And that from the knowledge of true form derives the true comprehension of what life is and what death is; and that, once the vain and puerile fear of death is quelled, we may know a part of the felicity that our contemplation affords, in keeping with the fundamentals of our philosophy, which withdraws the sombre veil of the insane belief in Orcus and in grasping Charon, a belief which poisons and detracts from all that is sweetest about our life. *Twelfth,* form is distinguished, not from the point of view of its substantiality, which forms its unity, but from that of the acts and the operations of its faculties, and from the point of view of the specific degrees of being that it produces. *Thirteenth,* we derive the genuine, definitive nature of the formal principle; how form is a perfect species, which is differentiated in matter according to the accidental dispositions that depend on the material form, inasmuch as this consists of diverse degrees and diverse dispositions of the active and passive qualities. We see how form is variable, and how it is invariable; how it defines and determines matter, and how it is defined and determined by matter. *Finally*, we show, through a certain comparison adapted for vulgar comprehension, how this form, this soul, can exist in its entirety in the whole and in any part whatsoever of the whole.

Argument of the third dialogue

In the third dialogue (after having, in the second, spoken of form, which has the nature of a cause more than that of a principle), we proceed to the examination of matter, which is thought to be more a principle or element than a cause. *First*, we show (not counting the prelude at the start of the dialogue) that David of Dinant was not led astray by taking matter to be an absolutely excellent and divine thing. *Second,* how, by different philosophical methods, we can give different definitions of matter, although there is, in reality, only one primary and absolute matter. Since it is manifested in different degrees, and is differently hidden under various species, different philosophers can understand it differently according to the definitions that suit them. It is no different for number, which is understood purely and simply by the arithmetician, harmonically by the musician, symbolically by the cabalist, and in still other ways by various wise men and fools. *Third,* the meaning of the word 'matter' is explained by means of the difference and the likeness that exists between the natural substratum and the artificial substratum. *Fourth,* we point out how the stubborn can be dispatched and to what extent we are obliged to meet their questions and argue with them. *Fifth,* from the true definition of matter it is inferred that no substantial form loses its being; and we forcefully prove that the Peripatetics and other vulgar philosophers have known no other substance than matter, even if they speak of the substantial form. *Sixth,* as a constant material principle is recognized, we demonstrate a constant formal principle; and we demonstrate that, from the fact of the diversity of dispositions that are in matter, the formal principle proceeds to the multiform configuration of different species and different individuals; and we show why it has come about that some, brought up in the Peripatetic school, have not wanted to recognize any other substance than matter. *Seventh,* why reason must distinguish matter from form, and potency from act; and we repeat what was stated in the second part concerning how we can, without laying ourselves open to criticism, grasp the substratum and the principle of natural things in diverse ways, according to different philosophical systems; more usefully, however, according to natural and magical methods, and more ineffectively according to rational and mathematical methods, especially if they submit so closely to the criterion and working of reason, that nothing worthwhile is produced in the end, nor any practical fruit gathered, without which all contemplation is to be reckoned futile.

Eighth, we present two points of view from which matter is generally considered: either as potency, or as substratum. And beginning with the first point of view, we differentiate matter in active potency and in passive potency, and in a certain way we guide it back to unity. *Ninth,* from the eighth proposition we deduce how what is supreme and divine is all that it can be, how the universe is all it can be, and how other things are not all that they can be. *Tenth,* as a result of what was said in section nine, we show in an estimable, clear and brief manner why there are vices, monsters, corruption and death in nature.

Eleventh, in what sense the universe is in none and in all of its parts – which occasions an excellent contemplation of divinity.

Twelfth, whence it happens that the intellect cannot grasp this absolute act and this absolute potency. *Thirteenth,* we conclude with the excellence of matter, which coincides with form as potency coincides with act. *Last,* from the fact that potency coincides with act, and that the universe is all that it can be, as well as for other reasons, we conclude that all is one.

Argument of the fourth dialogue

In the fourth dialogue (after having considered, in the third, matter insofar as it is a potency), matter is considered in so far as it is a substratum. We begin with Poliinnian distractions in order to present the definition of matter according to the vulgar principles of certain Platonists as well as of all the Peripatetics. *Second,* reasoning *iuxta* [according to] our own principles, we show that the matter of corporeal and incorporeal things is one, for several reasons, the first of which is drawn from the potency of one and the same genus. The second is drawn from a certain proportional analogy between the corporeal and the incorporeal, between the absolute and the contracted. The third is drawn from the hierarchy or ladder of nature, which goes up to a first embracing or comprehending principle. The fourth is taken from the fact that there must be something indistinct before matter is distinguished into corporeal and incorporeal: it is that indistinct which is represented by the supreme genus of the category. The fifth is taken from the fact that, since there is a common nature shared by the intelligible and the sensible, it must be the same for the substratum of sensibility. The sixth is drawn from the fact that the being of matter is independent of corporeal being, so that it is no less appropriate to incorporeal than to corporeal things. The seventh is derived from the hierarchy of the superior and inferior that is established between the substances; for where this hierarchy exists, we presuppose and understand a certain commonness in terms of matter, which is always signified by the genus, as the form is signified by the specific difference. The eighth derives from a principle alien to our philosophy but held by many, the ninth from the plurality of species that we attribute to the intelligible world. The tenth comes from the relation of similarity and imitation between the three worlds: metaphysical, physical and logical. The eleventh is drawn from the fact that all number, diversity, order, beauty and ornament are related to matter.

Third, we present briefly four opposing arguments and respond to them. *Fourth,* we show how this matter and that matter differ, how differently we convey this and that, and how matter coincides with act in incorporeal things, and how all the species of dimensions are in matter, all the qualities being comprised in form. *Fifth,* that no wise man has ever said that forms are received by matter as from outside, but that it is matter which, expelling them, so to speak, from its womb, produces them from within. It is therefore not a *prope nihil,* an almost nothing, a pure and naked potency, since all forms are contained in it, produced by it, and brought forth by virtue of the efficient cause (which, from the point of view of being, can even be indistinguishable from matter); they have no mode of actual existence in sensible and intelligible being other than through accidental existence, granted that all that which appears and is made manifest through the accidents founded on dimensions is pure accident, even if substance is always indivisible and always coincides with undivided matter. Hence, we see clearly that from explication we cannot get anything but accidents, and so the substantial differences are hidden, as Aristotle, checked by the truth, said. So that, pondering the subject well, we may conclude that the uniform substance is one, and that truth and being are one, which manifests itself through innumerable particularities and individuals, showing itself in countless, concrete, individual substances.

Sixth, how very far from all reason is what Aristotle and his like mean when they consider matter as being in potency, given that such a being is assuredly a nullity, since, according to them, matter is so absolutely permanent that it never varies or changes its being, all variation and modification being related to it, and since, still according to them, that which is, after having been able to be, is always composite. *Seventh,* we show how meaningless the characterization of matter as appetite is, using the self same logic derived from the principles and hypotheses of those very people who so strongly proclaim matter to be the daughter of privation and its appetite to be similar to the insatiable craving of an impassioned female.

Argument of the fifth dialogue

In the fifth dialogue, which deals chiefly with unity, the foundation of the edifice of all natural and divine cognition is laid. Here, *first,* we present the theme of the coincidence of matter and form, potency and act, so that being, logically divided into what it is and what it can be, is physically undivided, indistinct and one, and at the same time infinite, immobile and indivisible, with no difference between part and whole or principle and principled. *Second,* that in this one, there is no difference between a century and a year, a year and an instant, a palm and a stadium[3], a stadium and a parasang[4], and that in its essence this and that other specific being are not distinguished one from the other, because there is no number in the universe, and hence the universe is one. *Third,* that in the infinite, the point does not differ from the body, because there is no difference between potency and act; hence, if the point can extend in length, the line in breadth and the surface in depth, the point is long, the line broad and the surface deep; and all things are long, broad and deep, and therefore one and the same; and the universe is all centre and all circumference. *Fourth,* how Jove (as he is called), being found even more intimately in everything that the form of everything can be imagined to be (because he is the essence through which all that exists possesses being, and since he is in everything, each thing possesses the whole even more intimately than it does its own form), we may infer that all things are in each thing, and that, consequently, all is one. *Fifth,* we answer the sceptic who wishes to know why all particular things change, and why the particular matters, in order to receive this or that being, strive towards this or that form. We show how there is unity in the multiplicity, and multiplicity in the unity, how being is multimodal and multi-unitary, and how it is, finally, one in substance and in truth. *Sixth,* we deduce

[3] A unit of length, usually equal to 600 Greek or Roman feet, or one-eighth of a Roman mile.
[4] An Iranian unit of length, usually reckoned as equal to between 3 and 3½ miles (5 to 5½ km).

whence proceed that number and difference, as well as the fact that they are not being but of being and relative to being. *Seventh,* we show that whoever has discovered this one – I mean the essence of this unity – has uncovered the key without which one cannot enter into the true contemplation of nature. *Eighth,* by means of a new analysis, we reaffirm that the one, the infinite – that being, that which is in all – is everywhere, or better still, is itself the *ubique* [everywhere], and that, therefore, the infinite dimension, since it is not magnitude, coincides with the undivided individual, as the infinite multitude, since it is not number, coincides with unity. *Ninth,* how in the infinite there are no parts, however particularized the things of the universe are; where, consequently, all that we see of diversity and difference is nothing but diverse aspects of one and the same substance. *Tenth,* how in the two extremes that are assigned to the extremities of nature's ladder, we must see not two principles, but one only, not two beings, but one, not two contraries and opposites, but one and the same congruence. There height is depth, the abyss is inaccessible light, gloom is clarity, great is small, the confused is distinct, discord is amity, the divisible is indivisible, the atom is immensity – and all inversely. *Eleventh,* in what way certain geometrical terms such as point and unity may serve to lead us towards the contemplation of being and unity, although they are insufficient to express them. Whence Pythagoras, Parmenides and Plato should not be so foolishly interpreted according to Aristotle's pedantic criticism. *Twelfth,* from the fact that the substance or being is distinct from quantity, measure and number, we infer that it is one and undivided in all and in any thing whatsoever.

Thirteenth, we introduce the marks and the proofs that contraries indeed coincide, derive from the same principle, and form, in reality, but one substance: this is seen first mathematically, and then demonstrated physically.

Here, then, most illustrious Sir, you see where we must begin in order to venture towards a more specific and rightful cognition of things. It is there that (as within its exclusive seed) the host of natural science's conclusions is contained. Thence derive the structure, disposition and order of the speculative sciences. Without this introduction[5], all attempt, all exploration and all initiative are in vain. Pray accept, with a benevolent spirit, this principle, this one, this fountain, this wellhead, so that its descendants, its progeny, may be sparked to emerge, and so that its rivers and floods may flow forth more abundantly, and its numbers may continually multiply and members flourish; so that the night with its drowsy veil

[5] *Cause, Principle and Unity* was intended as an introduction to Bruno's *De l'infinito universo e mondi.*

and gloomy mantle may come to an end, allowing brilliant Titan, parent of the divine Muses, adorned with his family and surrounded by his everlasting court, to banish the nocturnal torches and brighten the world with a new day, surging forth again with his triumphant chariot from the vermilion bosom of this graceful Aurora.

Farewell

Giordano nolano
Al principi de l'universo

Lethaea undantem retinens ab origine campum
emigret, o Titan, et petat astra, precor.
Errantes stellae, spectate procedere in orbem
me geminum, si vos hoc reserastis iter.
Dent geminas somni portas laxarier usque,
vestrae per vacuum me properante vices:
obductum tenuitque diu quod tempus avarum,
mi liceat densis promere de tenebris.
Ad partum properare tuum, mens aegra, quid obstat,
seclo haec indigno sint tribuenda licet?
Umbrarum fluctu terras mergente, cacumen
adtolle in clarum, noster Olimpe, Iovem.

[from Giordano the Nolan to the Principles of the Universe

That the tenebrous Earth which, from the beginning, has cleaved to the wavy expanse of the waters, may leave its seat and fly towards the heavenly orbs, I beg you, O Sun. And you, wandering stars, behold me as I proceed towards the twofold heaven, since it is you who have opened this path to me. May your movements open before me, as I rush through the spaces, the doors of sleep: that which miserly time has kept long hidden, may it be allowed me to draw into the light out of the dense gloom. What prevents you, O suffering spirit, from hastening to give birth to your truth, though you bequeath it to an unworthy age? Though the flow of shadows submerges the Earth, you, my Olympus, make shine your peak in the clear heavens.]

Al proprio spirito

Mons, licet innixum tellus radicibus altis
te capiat, tendi vertice in astra vales.

Mens, cognata vocat summo de culmine rerum,
discrimen quo sis manibus atque Iovi.
Ne perdas hic iura tui fundoque recumbens
implicitus tingas nigri Acherontis aquas.
At mage sublimeis tentet natura recessus,
nam, tangente Deo, fervidus ignis eris.

[To his own spirit

O mount, though the Earth bounds you, holding you by the deep roots on which you repose, at the summit you can stretch to heaven. O mind, a sister mind from the high summit of the world calls you, to be the boundary between heaven and hell. Do not lose your rights here below, and do not touch the black waters of Acheron, falling to the bottom and becoming caught in it. Rather, investigate the sublime recesses of nature, since, if God moves you, you will become ardent fire.]

Al tempo

Lente senex, idemque celer, claudensque relaxans,
anne bonum quis te dixerit, anne malum?
Largus es, esque tenax: quae munera porrigis, aufers;
quique parens aderas, ipse peremptor ades;
visceribus educta tuis in viscera condis,
tu cui prompta sinu carpere fauce licet.
Omnia cumque facis cumque omnia destruis, hinc te
nonne bonum possem dicere, nonne malum?
Porro ubi tu diro rabidus frustraberis ictu,
falce minax illo tendere parce manus,
nulla ubi pressa Chaos atri vestigia parent
ne videare bonus, ne videare malus.

[To time

O old man, slow and swift, who opens and closes, must we speak well or ill of you? You are generous and stingy; the gifts you offer, you take back; you kill what you cause to be born, and what you generate from your bowels, in your bowels you devour, you to whom it is allowed to consume with your jaws the fruit of your bosom. You create all and destroy all: could I not, then, call you good, and call you evil? But when you will surprise me with your swift mortal blow, with your menacing scythe, let me stretch my hands forth to where there is no trace seen of black Chaos: thus, you will not appear good, nor appear bad.]

De l'amore

Amor, per cui tant'alto il ver discerno,
ch'apre le porte di diamante e nere,
per gli occhi entra il mio nume, e per vedere
nasce, vive, si nutre, ha regno eterno.
Fa scorger quant'ha il ciel terr' ed inferno,
fa presente d'absenti effigie vere,
repiglia forze, e, trando dritto, fere,
e impiaga sempre il cor, scuopre ogn'interno.
O dunque, volgo vile, al vero attendi,
porgi l'orecchio al mio dir non fallace,
apri, apri, se puoi, gli occhi, insano e bieco.
Fanciullo il credi, perché poco intendi;
Perché ratto ti cangi, ei par fugace;
Per esser orbo tu, lo chiami cieco.

[On love

Love grants me such a lofty vision of the truth that he makes the black doors of diamond open: through the eyes the god enters, and it is to see that he is born, lives, is fed, and reigns forever. He reveals all of heaven, hell and earth; makes appear true images of the absent; regathers strength to hit with a direct blow, always wounds the heart and reveals all that is hidden. Therefore, base mob, attend to the truth: lend your ears to my words, which do not deceive. Open, open if you can, your mad and squinting eyes. You call him a child, because you understand so little; because you are so inconstant, he seems fickle to you; your own sightlessness makes you call him blind.]

Untitled

Causa, principio, et uno sempiterno,
onde l'esser, la vita, il moto pende,
e a lungo, a largo e profondo si stende
quanto si dic'in ciel, terr'ed inferno;

> con senso, con raggion, con mente scerno
> ch'atto, misura e conto non comprende
> quel vigor, mole e numero, che tende
> oltr'ogn'inferior, mezzo e superno.
> Cieco error, tempo avaro, ria fortuna,
> sord'invidia, vil rabbia, iniquo zelo,
> crudo cor, empio ingegno, strano ardire
> non bastaranno a farmi l'aria bruna,
> non mi porrann'avanti gli occhi il velo,
> non faran mai ch'il mio bel sol non mire.

[O, you sempiternal cause, principle and one, whence depend being, life and movement, and whence in length, breadth and depth extends all that which is in heaven, on earth and in hell: with sense, reason and spirit I discern that act, measure and reckoning do not comprehend that force, mass and number which transcends all that is lowest, middle or highest. Blind error, greedy time, adverse fortune, deaf envy, vile rage, hostile zeal, cruel hearts, perverse spirits, bizarre passions will not suffice to obscure the air before me, nor place the veil before my eyes, nor ever stop me from beholding my beautiful sun.]

Giordano Bruno, Nolan
Cause, principle and unity

First dialogue

Speakers: Elitropio, Filoteo, Armesso

ELITROPIO. Like felons used to the darkness, who come up to the light when freed from the depths of some gloomy tower, many trained in common philosophy, and others, will be clutched by fear, seized with astonishment and (unable to stand the new sun of your shining concepts) thoroughly unsettled.

FILOTEO. It is not the fault of the light, but of their sight: the more excellent and beautiful the sun, the more hateful and harshly unwelcome it will be to night-witches' eyes.

ELITROPIO. In your hope of raising us out of the blind abyss, into the sight of the open, peaceful and tranquil stars that shine with such beautiful variety against the cerulean mantle of heaven, Filoteo, you have picked an uncommon, unusual and difficult venture. And though the helping hand of your compassion is held out to us men, the ungrateful will still attack you in ways that are as varied as the many animals generated and nourished within the gentle earth's ample and maternal bosom; for it is clear that the human species displays, in the particularities of its individuals, the variety of all other species together. In each of our individuals, the whole is present more explicitly than in the individuals of other species. Thus, some, as soon as they feel the fresh air, like the bleary-eyed mole, will tunnel straight back down into the earth to seek their natural, inky depths. Others, like night birds, on seeing the vermilion ambassadress of the sun come up in the east, will be forced by the weakness of their eyes to repair to their dingy retreats. All creatures banished from the presence of the celestial lights and doomed to the eternal chasms, cages and caverns of Pluto – all animals, called by the horn of the fearsome Erynnis, Alecto, will spread their wings and flee headlong to their dwellings. But the animals born to behold the sun, having waited out the hated night, will give thanks to the merciful heavens and, prepared to gain within the globose crystals of their eyes the rays for which they have so long waited and pined, will adore the east, not only with unwonted adoration in their hearts, but with voices and hands. Men will begin to speak when from the east's gilded balcony, handsome Titan has let loose the fiery steeds who cleave the sleepy silence of the moist night. The docile, defenceless and simple sheep flocks will bleat; the horned oxen will bellow, heeded by their rustic ox-herders; and Silenus' quadrupeds will start to bray, frightening the stupid Giants again for the gods' benefit. Tossing in their muddy beds, the boars will deafen us with their obstinate grunting. Tigers, bears, lions, wolves and the cunning fox poking his head from the cave will behold from their high deserts their flat hunting grounds, and will let forth from ferocious breasts their roars, growls, snarls, howls and cries. In the air and on the fronds of branchy trees, the roosters, eagles, peacocks, cranes, doves, blackbirds, crows, sparrows, nightingales, magpies, ravens, cuckoos and cicadas will lose no time responding, re-echoing with their earsplitting chatter. Still further, from their mobile, liquid dominions, the white swans, the many-hued water fowl, the swift razor-bills, the marsh ducks, the honking geese and the carping frogs will disturb our ears with their din, such that the warm sunlight diffused in the air of our privileged hemisphere will find itself attended, greeted and perhaps plagued by cries as numerous and as varied as are the breaths that drive them out from the hollows of their respective breasts.

FILOTEO. It is not only common, but necessary and natural that every animal utters its own cry. Beasts cannot form regulated accents and articulated sounds like men, since their physical makeup, nourishment and tastes are dissimilar.

ARMESSO. Please give me a chance to speak also: not about light, but of some circumstances that, far from comforting the senses, injure the feelings of whoever observes and reflects. For your own peace and quiet (that I wish for you with fraternal affection), I would not want these speeches of yours to be made into comedies, tragedies, laments, dialogues, or what have you, like the ones that circulated openly a while ago, and forced you to stay shut up in your homes.

FILOTEO. Speak frankly.

ARMESSO. I have no intention of speaking like a holy prophet, as an abstruse oracle, like an apocalyptic visionary or the she-ass of Balaam[1] beholding the angel. Nor will I discourse as if I were exhilarated by Bacchus or swollen with wind by the sluttish Parnassian muses, nor like a Sibyl impregnated by Phoebus, nor like a prognostic Cassandra,[2] nor as if Apollonian rapture had seized me from my toenails to the hair on my head, nor like the seer illuminated in the oracle or Delphic tripod, nor like wise Oedipus,[3] probed in the riddles of the Sphinx, nor as a Solomon before the enigmas of the queen of Sheba,[4] nor like Calchas,[5] interpreter for the Olympian council, nor as a Merlin possessed, nor as one emerged from the cave of Trophonius.[6] Instead, I will speak in common, vulgar language, like a man who has had other things on his mind than to go about distilling the juice of his brain and cerebellum to the point of withering his *pia mater* and *dura mater*. What I mean is that I will talk as one who has no wits but his own, and to whom not even the garden- or kitchen-variety gods among the celestial court condescend to cast a straw, though they heap their favours *ad infinitum* even on their horses – those gods, I say, that ordinarily show themselves more intimate, more familiar and congenial with us. I mean Bacchus, or the drunk mounted on the ass, or Pan, or Vertumnus, or Faunus, or Priapus: the ones who neither drink ambrosia nor taste nectar (unappreciative of nymphs and pure water), but quench their thirst at the bottom of the barrel with sour wines.

ELITROPIO. Too long a preface.

ARMESSO. Patience: the conclusion's swift. To put an end to this, I would like to say that I will offer you words that need no deciphering, as if they had been distilled, passed through an alembic, condensed in a double-boiler

[1] Numbers 28. [2] Virgil, *Aeneid*, II, 246–7. [3] Seneca, *Oedipus*, vv. 92–3 and 101–2.
[4] 1 Kings, 10, 1–3. [5] Virgil, *Aeneid*, II, 122–4. [6] See Erasmus, *In Praise of Folly*, I.

and sublimated with a prescription of the quintessence, but such words as my wetnurse hammered into my skull – a woman as thick-skinned, big-chested, wide-hipped, ample-bellied and broad-bottomed as that Londoner I caught sight of in Westminister, who possessed such ample mammeries, like hot water bottles for her stomach, that they seemed the halfboots of the immense Saint Paragorio, and which if tanned would match a pair of Ferrarese bagpipes.

ELITROPIO. That is quite enough for a preface.

ARMESSO. Well, then, to come to the rest – leaving a little to one side observations and opinions concerning light and the potential splendor of your philosophy – I would like to hear from you in what terms you wish us to greet, in particular, that brilliant doctrine which shines forth from *The Ash Wednesday Supper.* What animals are those that perform in *The Ash Wednesday Supper?* Are they aquatic, aerial, earthly or lunatic? And, leaving aside the observations of Smitho, Prudenzio and Frulla,[7] I would like to know if they are wrong or right, who claim that you bark like a rabid dog, in addition to sometimes playing the monkey, sometimes the wolf, sometimes the magpie, sometimes the parrot, now this animal, now that, mixing grave and serious words, moral and natural, ignoble and noble, philosophic and comic.

FILOTEO. Do not be surprised, brother, for the scene was indeed a supper, where brains are guided by the passions aroused by the flavours and odours of food and drink. The supper will be verbal and spiritual as a logical consequence of its material and corporeal guise. The dialogue has parts as different and various, therefore, as those which ordinarily comprise that other supper. The former has conditions, circumstances, and means of its own that are as peculiar to it as those of the latter may be.

ARMESSO. Help me get at your meaning, please.

FILOTEO. In the one case (as is fitting and proper), there are the salads and main dishes, fruits and common victuals, hors d'oeuvres and spices, warm and cold, raw and cooked, food of aquatic and terrestrial origin, cultivated and wild, ripe and green, food for the healthy and for the ill, dishes for gourmets and dishes for the hungry, ones that are light and substantial, bland and salted, tart and sweet, bitter and mild. Similarly, in the other case and by a certain analogy, contradictions and differences have appeared, suitable to the various stomachs and tastes of those whom it may please to take part in our symbolic banquet, so that no one can complain

[7] Secondary characters in *The Ash Wednesday Supper.*

of having attended in vain, and whoever does not like one thing can help himself to another.

ARMESSO. True. But what is your answer if, in addition, at your banquet, at your supper, things appear that are good neither for salads nor main dishes, nor good as fruits or fillers, neither cold nor hot, raw nor cooked, good for neither the healthy nor the sick, stuff that should never have left the hands of the cook or confectioner, stuff that stirs no appetite and satisfies no hunger?

FILOTEO. You will see that, in this, our supper does not differ from any other that may be dished out. As with that other, as you are savouring your meal, you might scald your mouth with a bite that is too hot, so you have to either spew it back out or bandy it about your palate with tears and crying until you can give it that cursed shove in order to speed it down your gorge, or you jar some tooth, or you bite into your tongue at the same time as your bread, or else there is a piece of grit that breaks up and sticks between your teeth, forcing you to spit out the whole mouthful. Perhaps some hair or whisker off the cook glues to your palate and makes you nearly vomit, or else a fishbone lodges in your gullet and makes you wheeze, or another small bone lodged sideways in your throat threatens to suffocate you. To our and to everybody's displeasure, analogous and equivalent things have been found at our supper. All that is due to the sin of the first man, Adam. Because of our ancestor, perverse human nature is condemned to find disgust joined to delight.

ARMESSO. Spoken with sanctity and piety. But what is your answer to those who call you a raging cynic?

FILOTEO. I will concede the point readily, at least in part.

ARMESSO. But you know that it is less dishonourable for a man to undergo abuses than to inflict them?

FILOTEO. Yet it is enough that my actions are labelled vengeance, and the others' abuse.

ARMESSO. Even the gods are liable to receive insults, suffer censure, and bear reproach; but to insult, censure and reproach are the business of people who are low, mean, cowardly and worthless.

FILOTEO. True. That is why we do not injure, but rather rebutt the injuries that are cast, not so much at us but at condemned philosophy, proceeding in such a way that other insults are not added to those already received.

ARMESSO. So you want to act the biting dog, so that no one dares molest you?

FILOTEO. Exactly, because I desire peace, and unpleasantness dipleases me.

ARMESSO. Yes, but they deem that you proceed with too much severity.

FILOTEO. That is to prevent them from coming back, and to daunt others from coming to dispute with me or someone else, and treating our demonstrations with such expedients.

ARMESSO. The offence was private, the retaliation public.

FILOTEO. None the less just for that. Many errors are committed in private, which are then justly chastised in public.

ARMESSO. But that way you end up ruining your reputation, and making yourself more blameworthy than those others, because the world will declare you impatient, fantastic, daft and bizarre.

FILOTEO. It does not matter, as long as they stop harassing me from now on, and if I shake the club of the cynic at them, that is so they will let me go about my business in peace. It is clear they do not want to do me kindnesses, but they should not exercise their coarseness on me.

ARMESSO. But do you think it is fitting for a philosopher to set about avenging himself?

FILOTEO. If those who harassed me were a Xanthippes, I would be a Socrates.

ARMESSO. Don't you know that patience and long-suffering does everyone good, and that through them we become like heroes and celebrated gods, who, according to some, defer their vengeance, and according to others, neither take revenge nor give way to anger?

FILOTEO. You are wrong to think I cared to have revenge.

ARMESSO. What then?

FILOTEO. I was concerned with correcting – an activity that also makes us similar to gods. You know that Jove ordered poor Vulcan to work even on holidays, so that his cursed anvil goes on receiving the fierce blows of the hammer eternally. No sooner is one raised than another comes smashing down, so that the righteous thunderbolts used to chastise the guilty and unlawful are never in short supply.

ARMESSO. There is a difference between you and Jove's blacksmith, husband of the goddess of Cyprus.

FILOTEO. It is enough, in any case, that I am perhaps not unlike the gods in patience and long-suffering. And those qualities were put to the test in this affair, in that I never gave full rein to my scorn and never spurred on my anger.

ARMESSO. To be castigator – of the multitude, especially – is not the job of just anyone.

FILOTEO. Add to that: especially when he has nothing to do with the multitude.

ARMESSO. They say that you must not be an agitator in a country not your own.

FILOTEO. And I say two things: first, one should not kill a foreign physician because he tries to administer cures not administered by the natives; second, I say that the true philosopher's country is all the world.

ARMESSO. But if they consider you neither philosopher nor physician nor countryman?

FILOTEO. That does not make me any the less so.

ARMESSO. Who will vouch you that?

FILOTEO. The gods who have put me here, I who find myself here, and those who have eyes to see me here.

ARMESSO. Your witnesses are very few and little-known.

FILOTEO. The true physicians are indeed few and little-known, while all these people are truly ill. And I repeat that they have no right to inflict or to allow others to inflict such treatment on those who offer honest merchandise, foreigners or not.

ARMESSO. Few are acquainted with this merchandise.

FILOTEO. Pearls are no less precious for that reason, nor do we, therefore, dedicate less effort rallying to their defence, to save and vindicate them with all our might from the trampling of swine. May the gods favour me, Armesso, since I have never carried out acts of vengeance out of sordid self-love or low self-interest, but out of devotion for the offended majesty of my belovèd mother, philosophy. False friends and false children (for there is no worthless pedant, do-nothing phrasemaker, stupid faun or ignorant hack who does not aspire to be numbered among her family by showing up loaded with books, growing out his beard, or getting up prosopopoeical by other means) have wasted her so away that, among the common people, philosopher rhymes with impostor, quack, swindler, good-for-nothing, charlatan and howling pedant, good only as home entertainment or country scarecrow.

ELITROPIO. Indeed, philosophers as a race are rated by most men as more despicable than house chaplains sprung up from the dregs of humanity, who, however, disgrace the priesthood far less than the philosophers, chosen from among every sort of beast, have shamed philosophy.

FILOTEO. So let us praise the ancient race. Philosophers then had so much value that from their ranks were recruited lawmakers, counsellors and kings. And the counsellors and kings were such that from those functions they were elevated to the priesthood. In our age, most of the priests are such that they themselves are discredited, and do discredit to the divine laws; nearly all the philosophers we see are worth so little that they are disparaged along with their science. What is worse, a multitude of scoundrels, like a mass of nettles, have grown used to smothering with poisonous mirages what little truth and virtue get revealed to the few.

ARMESSO. I know no philosopher who gets so aroused in favour of discredited philosophy, nor do I perceive any, Elitropio, as impassioned by his science as Teofilo. What would happen if all other philosophers had the same character, I mean if they had so little patience?

ELITROPIO. Those others have not made so many discoveries, nor do they have as much to preserve or defend. They can easily devalue philosophy that is worthless, or what is nearly worthless, or that which they do not know; but one who has found truth, which is a hidden treasure, is inspired by the beauty of that divine face and grows jealous to defend her against plunder, negligence and contamination. Just so, a miser may conceive a passion for gold, diamonds and carbuncles, or a man for the beauty of a foul woman.

ARMESSO. But let us get back to our subject and arrive at the *quia* [why]. They say of you, Teofilo, that in your supper you criticize and insult a whole city, an entire province, a complete kingdom.

FILOTEO. That, I never thought, never intended, never did. If I had ever thought, wished or done so, I would condemn myself with utmost severity and bend over backwards to make a thousand disavowals, retractions and disclaimers; not only if I had insulted a noble and ancient realm such as this, but any other, however great its reputation for barbarism. And I mean not only if I had offended any city, however widespread its reputation for incivility, but even if I had insulted any class whatsoever, however savage it was held to be, or even any one family, however inhospitable it was considered. There cannot be a race, kingdom, city or house where contrary and opposing manners do not exist, and to which one can assign the same temperament to all, such that it is impossible for one man to find pleasure in what displeases another.

ARMESSO. As far as I am concerned, I have read, re-read and meditated upon all you have said (although on some points, I do not know just why, I find you a bit excessive), and you seem to me for the most part to proceed with moderation, reason and discernment; but the noise has spread as I have set out.

ELITROPIO. That noise of this and other things has been bandied about through the meanness of some of those who felt themselves touched. Eager to take revenge, but conscious of the weaknesses of their arguments, their doctrine, their intelligence and their strength, they not only fabricate as many lies as they can, to which no one but their like gives credit, but they try to enlist partisans by making out that your condemnation of some individuals constitutes a pervasive insult.

ARMESSO. I think, on the contrary, there are people, not without wisdom and judgement, who gauge the insult universal because you indicate certain manners as belonging to people of this or that nation.

FILOTEO. But what are these alleged manners? Are not similar or worse ones, not to mention manners much more peculiar in genus, species and number, found in the most excellent parts of the world? Would you claim that I were abusive and ungrateful toward my own country, if I said that in Italy, in Naples or Nola, similar or more criminal manners can be found? Would you say that I had abused that blessèd realm, often set at the head and the right hand of our globe simultaneously, governor and tamer of the other nations (and ever regarded by us and by others as mistress, nurse and mother of all the virtues, disciplines, humanities and the qualities of modesty and courtesy), when esteemed poets, themselves, have justly sung its praises, but yet do not shrink from calling her, if the occasion requires, mistress of all vice, error, greed and cruelty?

ELITROPIO. This is certainly in keeping with the precepts of your philosophy, by virtue of which you maintain that contraries coincide both in principle and in reality. Thus, minds most suited to high, worthy and generous enterprises will fall, if they are perverted, into extreme vice. Moreover, we generally find the rarest and choicest wits amongst the most foolish and ignorant folk, and there where the people are generally the least civil and the most lacking in courtesy, we find, in some individual cases, extreme civility and good manners – so that, in one way or another, many nations seem to have received an equal measure of perfections and imperfections.

FILOTEO. What you say is true.

ARMESSO. And yet, Teofilo, I am distressed, as are many others, that in our friendly nation you have come up against the kind of people who have so irked you that you vent your complaints by means of a sooty supper, instead of having met those, much more numerous, who would have shown you how much our country (even if it is presented by your countrymen as *penitus toto divisus ab orbe*[8] [utterly cut off from the whole world]) is disposed to all literature, arms, chivalry, humanities and courtesy. We venture with all our strength not to be inferior to our ancestors in those domains, nor to be outclassed by other nations – especially those who believe themselves naturally endowed with noble manners, science, arms and civility.

FILOTEO. On my faith, Armesso, I neither would nor could contradict anything you say, neither with words, nor with reasonings, nor in conscience. You defend your cause with extreme modesty and keen argument, rather than attacking me out of some sort of barbarous pride. Thus I deplore all the more the fact that the individuals of whom we have been speaking have given me occasion to pain you, and others of honourable and humane temperament. I am beginning to feel sorry that those dialogues were ever published, and, if it will please you, I will see to it that they are circulated as little as possible.

ARMESSO. My pain, like that of other very noble souls, stems so little from the publication of those dialogues that I would willingly undertake to have them translated into our tongue, in order to serve as a lesson for those few among us who are so lacking in education and manners. Maybe, on seeing with what nerve their impertinent attacks are received and how inappropriate they are, and with what traits they are described, even if they choose not to change tack and follow the examples and the lessons of the best and brightest men, they might at least amend their ways and imitate them out of the shame of being identified as part of that number. They might learn that honour and courage are not forged by the capacity and the art of molesting but by quite opposite behaviour.

ELITROPIO. You show much ability and shrewdness in defence of your country, and in contrast to the crowd of those poor in arguments and wisdom, you know how to recognize and appreciate others' merits. But Filoteo does not seem to me as deft in defending himself and protecting his reputation. As nobility and rusticity differ, just so opposing effects are to be expected and feared from them. On one hand, a Scythian oaf will manage to look wise and will be celebrated for his success if, leaving the banks of the Danube, he goes away, bearer of audacious reproaches and legitimate complaints, to put to the test the authority and majesty of the Roman

[8] Virgil, *Bucolics*, I, 66: 'et penitus tot divisos orbe Britannos' ('(the Britons) isolated at the end of the world').

Senate, which, if it finds in his censure and invective occasion to accomplish an act of high prudence and magnanimity, does its severe critic the honour of a colossal statue. On the other hand, a Roman senator and gentleman would demonstrate very scarce wisdom in abandoning the mild banks of the Tiber, even armed with legitimate complaint and completely justified reprimand, to go try the Scythian oafs, who would seize the occasion to build, at his expense, towers and Babels of arguments of the utmost baseness, insolence and infamy, unleashing popular fury and stoning him in order to show other nations how much difference there is between dealing with human beings and with those who are merely made in their image and likeness.

ARMESSO. Let it never come to pass, Teofilo, that I could or should consider it proper for me, or anyone else endowed with even greater judgement than myself, to take up the cause of those who are the object of your satire under the pretext that they are of our nation, which some natural law impels us to defend. I will never admit – nor could I ever be anything but the enemy of anyone who makes such a claim – those people as countrymen. Our nation is comprised exclusively of people as noble, civil, polite, educated, measured, humane and reasonable as those of any other place. Even if such people exist within our borders, surely they are nothing but filth, scum, dirt and swine; part of the kingdom, or city, only in the sense that the bilge is part of a ship. We should not, therefore, bother ourselves overmuch about such individuals, because in doing so we might grow as injurious as they are. Among their ranks I include numerous priests and doctors, some of whom certainly become gentlemen, thanks to their doctorates. But most of them, who before did not dare show their rude authority, come boldly and arrogantly out into the open, later becoming hardier and more presumptuous when they rise to the titles of literary men and priests. Hence, it is no wonder that you see swarms of those who, despite their priesthood and their doctorate, retain more of the herd, the flock and the stable than actual ploughmen, goatherds and grooms. Thus, I would have preferred you had not attacked our university so harshly, condemning it as a whole, so to speak, without regard for what it once was, and can or will be in future, and is, in part, today.

FILOTEO. Have no fear. Although on this occasion we looked primarily at your university, it commits no worse errors than others whose members consider their academy superior, but which produce asses dressed up with diadems and hacks decked with rings under the title of doctors, for the most part. However, I do not dispute the great value of your university's original statutes, nor the beauty of its programme of studies, nor the majesty of its ceremonies, nor the fine organization of its works, nor the solemnity of its traditions, not to mention other qualities which serve to honour and embellish any university, and for which it must doubtless be considered the finest in Europe and, therefore, the world. And I cannnot deny that, as far as fineness of spirit and sharpness of wit are concerned, both of which Britain produces naturally here and there, your university really is similar to, and may be on par with, the best schools elsewhere. We have not forgotten, either, that speculative studies first flourished here, before spreading to other parts of Europe, nor that its princes of metaphysics (though barbarous of tongue and cowled by profession) have disseminated the splendour of a most rare and noble part of philosophy, in our day nearly extinct, to all the universities of non-barbarous countries. But one thing concerns me that seems annoying and comical at the same time. Although I have not found doctors more Roman and more Attic than these here, still, for the most part, they boast that they are the opposites of their forerunners, resembling them in nothing – those predecessors who, caring little for eloquence or grammatical rigour, devoted themselves entirely to speculative research, called by these current doctors 'sophismata'. As for myself, I prize the metaphysics of these latter more, in which they surpassed their teacher Aristotle – notwithstanding the fact that it is impure, and dirtied with certain empty arguments and theorems that are neither philosophical nor theological, but the products of idle or badly-used intellects – than what the others today can bring us, with all their eloquence and Ciceronian declamatory art.

ARMESSO. Those arts are not to be belittled.

FILOTEO. True, but if we have to choose between the two, I set the culture of the mind, however mean it may be, over that of words and phrases, however eloquent.

ELITROPIO. Your comment brings Fra Ventura to mind. Commenting on the Gospel passage *'reddite quae sunt Caesaris Caesari'*[9] [render unto Caesar that which is Caesar's], he cites on that occasion the names of all the coins circulating at the time of the Romans, with their imprints and their weights – names that he had come across in I do not know what damned annals or opuscules, more than a hundred and twenty of them – in order to show us the range of his studies and the power of his memory. At the end

[9] Matthew 22, 21.

of his sermon some fine fellow accosted him and said, 'Reverend father, lend me a carline.' To which he answered that he belonged to an order of mendicants.

ARMESSO. What's the point of this story?

ELITROPIO. I mean that those who are versed in the science of names and phrases but do not worry about things are astride the same ass as that reverend father of asses.

ARMESSO. I think that, apart from the study of eloquence, in which they outshine all their predecessors and are unsurpassed by their contemporaries, they are neither destitute in philosophy nor in other speculative disciplines. Without ability in these, they cannot be promoted to any rank, because the university statutes, to which they are bound by oath, resolve that '*Nullus ad philosophiae et Theologiae magisterium et doctoratum promoveatur, nisi epotaverit e fonte Aristotelis*' [Let no-one who has not drunk of the Aristotelian fountain be promoted to the title of master and doctor of philosophy and theology].[10]

ELITROPIO. Ah, but I will tell you what they have done to avoid perjuring themselves. To one of the three fountains of the university they have given the name *Fons Aristotelis* [Aristotelian fountain], they have called another *Fons Pythagorae* [Pythagorean fountain], and the third is dubbed *Fons Platonis* [Platonic fountain]. Since the water to make beer and ale is drawn from these three fountains, as well as the water for horses and cows, it follows that nobody who has spent three or four days in those study rooms or colleges fails to drink, not only of the Aristotelian fountain, but also of the Pythagorean and Platonic.

ARMESSO. Too true, unfortunately. So it happens, Teofilo, that doctors come as cheaply as sardines, since they are made, found and hooked with little trouble. The herd of doctors today being thus (leaving aside the reputation of some of them, such as Tobias Matthew,[11] Culpepper,[12] and others whose names I have forgotten, distinguished alike for their eloquence, their doctrine and their high courtesy), the result is that the title of doctor, far from crediting one with a supplementary degree of nobility, places one under

[10] No such formula is found in the *Statuta antiqua Universitatis Oxoniensis*, ed. Gibson (Oxford, 1931). Aristotelian doctrines are, however, stressed in the statutes.

[11] Tobias Matthew (1546–1628), President of St John's College from 1572 to 1577, Dean of Christ Church from 1576 to 1584, and Vice-Chancellor of Oxford in 1579. Promoted to Dean of Durham in 1583 and named Bishop of York in 1606.

[12] Martin Culpepper, Professor of Medicine and Rector of New College from 1573 to 1599, Dean of Chichester from 1577, and Vice-Chancellor of Oxford in 1578.

the suspicion (unless one is known personally) of having a completely opposite nature and character. Hence, it happens that even men noble by birth or by accident, and enriched by the principal part of nobility which is learning, are ashamed to be promoted to the title of doctor, and so content themselves with merely being learned. You will find many more of these in the courts than you will among the pedants at the university.

FILOTEO. You will find both kinds everywhere there are doctors and priests, Armesso, so hold your complaining. Those who are true doctors and true priests, even if of modest origin, can only gain in civility and nobility, because knowledge is the most expedient way of making the human soul heroic. The more those others thunder from on high with *divum pater* [divine father], like the giant Salmoneus,[13] the more clearly they reveal their rudeness, strutting like satyrs or fauns dressed in purple, with that horrendous and imperial majesty, after having determined from the height of their magisterial chair to what declension *hic* [this, masc.], *haec* [this, fem.] and *hoc nihil* [this, nothing] belong.

ARMESSO. Let us change the subject. What is that book in your hand?

FILOTEO. Some dialogues.

ARMESSO. The *Supper?*

FILOTEO. No.

ARMESSO. What, then?

FILOTEO. Others where the themes of cause, principle and unity are treated according to our system.

ARMESSO. Who are the speakers? Are there, by any chance, some other devils in it like Frulla or Prudenzio, who will land us into trouble again?

FILOTEO. Rest assured that, except for one of them, they are all very peaceable, honest subjects.

ARMESSO. So that from what you say we will still have to pick some thorns out of these dialogues?

FILOTEO. No doubt. But you will be scratched where it itches, instead of pricked where it hurts.

ARMESSO. What else?

FILOTEO. Here you will meet, as first speaker, that erudite, honest, affable, polite and faithful friend Alexander Dicsono, who proposes the subject of the debate, and whom the Nolan loves as his own eyes. He is introduced as the one who furnishes Teofilo with his subject. Then Teofilo (who is myself) comes second, profiting by the occasion to make

13 See Virgil, *Aeneid*, 585–6.

distinctions, give definitions and carry out demonstrations concerning the theme proposed. Thirdly, you have Gervasio, not a philosopher by profession, but who likes to pass the time by attending our discussions; a person of indifferent odour who finds everything Poliinnio does comic, and from time to time gives him full rein to express his folly. The latter sacrilegious pedant is the fourth speaker; being one of those stern censors of philosophers, he claims to be a Momus, passionately attached to his flock of students, reputed to be a follower of Socratic love, an eternal enemy of the female sex. He considers himself, therefore, in order not to seem involved with physics, an Orpheus, a Musaeus, a Tityros, an Amphion. He is one of those who, when they have put together a beautiful conceit, composed an elegant little epistle or made off with a nice phrase from the Ciceronian kitchen, are at once Demosthenes come back to life, Tullius rejuvenated, Sallust who lives again, or an Argus who makes out every letter, every syllable and every word. He is Rhadamanthus who *umbras vocat ille silentum* [calls the shadows of the silent], or the Cretan king Minos who *urnam movet* [shakes the drawing-urn].[14] He is one of those men who puts every word to the test, and who mounts a debate around every phrase, saying that these are poetic, these sound comic, these are oratic; this is sweet, that is sublime, this other one is *humile dicendi genus*[15] [humble oratory genre]; this harangue is harsh, it would be lighter if composed like this, such and such a writer is not eloquent, he is little read in the ancients, *non redolet Arpinatum, desipit Latium*[16] [he does not smack of Arpinum, he lacks knowledge of Latin]. This word is not Tuscan, neither Boccaccio, nor Petrarch, nor other approved authors use it. One should write 'omo' and not 'homo', not 'honour' but 'onour', 'Poliinnio' instead of 'Polihimnio'. This kind of thing fills him with triumph, self-satisfaction and utmost pleasure with whatever he does. He feels himself a Jove who, from his high perch, gazes down on and contemplates the lives of other men, subject to so many errors, calamities, miseries and vain strivings. He alone is happy, only he lives a heavenly life, when he contemplates his divinity in the mirror of a *Spicilegium*,[17] a *Dictionarium*, a

[14] See Virgil, *Aeneid*, VI, 432–3: 'quaesitor Minos urnam movet; ille silentium/consiliumque vocat …' ('Minos reigned as the presiding judge, moving the drawing-urn, and called a jury of the silent ones …').

[15] *Humile dicendi genus* is the first of the three *genera dicendi* which Cicero distinguishes in his *tripertita varietas* of styles of oration.

[16] Cicero, born in Arpinium, sixty miles south-east of Rome. See the anti-Ciceronian satire of Erasmus, the *Ciceronianus* (1528).

[17] Title of a work by L.G. Scoppa, the *Spicilegium*, which dates from 1511.

Calepino,[18] a *lexicon*, a *Cornucopiae*,[19] a *Nizzolio*.[20] Endowed with such self-sufficiency, he alone is everything, while each of us is but one. If he happens to laugh, he calls himself Democritus; if he by chance cries, he calls himself Heraclitus; if he argues, he baptises himself Chrysippus; if he reasons, his name is Aristotle; if he forges chimeras, he becomes Plato; if he bellows out some paltry speech, he is Demosthenes; if he expounds Virgil, he is Maro. So he corrects Achilles, approves Aeneas, reprehends Hector, exclaims against Pyrrhus, laments Priam, accuses Turnus, excuses Dido, praises Achates, and finally, while *verbum verbo reddit* [he translates word for word], chaining together his barbarous synonyms, *nihil divinum a se alienum putat* [he maintains that nothing divine is alien to him]. He then descends haughtily from his chair, as if he had put the heavens in order, reformed worlds, organized senates and tamed armies. He is sure that, if it were not for the injustice of the times, he would convert into action what he has accomplished in thought. *O tempora, o mores!*[21] [O age, o manners!] How rare are they who understand the nature of participles, adverbs and conjunctions! How much time has flowed by without discovering the reason, the true cause, that makes the adjective agree with the noun, the relative join together with the antecedent, and the rule which places it at the beginning or end of a sentence, and the frequency and order in which one must slip in those interjections *dolentis* and *gaudentis* [of pain and joy], *'heu'*, *'oh'*, *'ahi'*, *'ah'*, *'hem'*, *'ohe'*, *'hui'* and other seasonings, without which the whole discourse is totally bland!

ELITROPIO. You can say whatever you like, and think as you wish, but I hold that in order to be happy in this life, it is better to imagine oneself Croesus, and be poor, than to imagine onself poor, and be Croesus. Is it not more conducive to beatitude to have a slattern you think beautiful and who satisfies you, rather than a Leda or a Helen who bores you and whom you end up abandoning? What does it matter, then, to those people, whether they are ignorant and ignobly occupied, when their happiness is in direct proportion to their own self-esteem? The ass likes fresh grass and the horse barley, just the same as you who like white bread and partridge; the hog is

18 Name of the lexographer Ambrogio Calepino, whose *Dictionarium* (which appeared before 1510) was so often reprinted during the 16th century that 'calepino' became synonymous with 'dictionary'.

19 Literally, 'horn of plenty', an allusion to Nicolò Perotti, *Cornucopiae sive commentaria linguae latinae* (Venice, 1489) and often reprinted during the 15th and 16th centuries.

20 *Nizzolio,* a synonym during the 16th century for 'Ciceronian lexicon'; See Mario Nizzoli, *Observationum in M.T. Ciceronem Prima [Secunda] pars,* (1535), reprinted several times.

21 Cicero, *Catilinam,* I, 2.

as happy with his acorns and slops as Jupiter with nectar and ambrosia. Do you want, by chance, to disabuse them of their agreeable folly when, in return for the cure, they come and break your head? I will leave aside the question of which is folly: the illusion, or its cure. A Pyrrhonist once said, 'Who knows whether our state is not death, and that of the alleged dead, life?' Who knows if true happiness and true beatitude do not consist of the due linking and taking apart the parts of a phrase?

ARMESSO. The world is such that we play Democritus at the expense of the pedants and grammarians, and diligent courtiers play at being Democritus at ours, while unthinking monks and priests democratize at everybody's expense. The pedants mock us, give-and-take, we sneer at the courtiers, and everybody at the monks. The outcome is that, since one is a fool in the eyes of the other, we are all fools, differing by species, but concordant *in genere et numero et casu* [in their genus, number and case].

FILOTEO. Just so censure differs in manner, kind and degree. Yet we must bend our knees and bow our heads before that most harsh, severe, horrendous and frightening censure of our arch-pedagogues. It is towards them we must turn our gaze and lift our hands, sighing, calling out, weeping and begging for mercy. Thus, it is to you that I turn, to you, who hold in your hand the caduceus of Mercury in order to resolve controversies; to you, who settle the differences that arise between men and gods. You, Menippos, who, from your seats on the moon's globe, look down on us with narrowed eyes from on high, noting our actions with repugnance and scorn. You, shield-bearers of Pallas, standard-bearers of Minerva, Mercury's stewards; you, Jupiter's custodians, Apollo's milk brothers, Epimetheus' co-thieves, Bacchus' bottlers, Euhan-criers' horse-grooms; you, who scourge the Edonides, spur on the Thyiades, excite the Maenads, seduce the Bassarids; you, the riders of the Mimallonides, copulators of the Egerian nymph, moderators of enthusiasm, demagogues of wandering peoples, decipherers of the Demogorgon, Dioscures of fluctuating disciplines, treasurers of the Pantamorpheus and bullock-emissaries of the highpriest Aron: to you we recommend our prose, submit our Muses, our premises, subsumptions, digressions, parentheses, applications, clauses, periods, constructions, adjectives and epithets. O you, sugarwater vendors, who ravish our spirits with your sweet little refinements, binding fast our hearts, fascinating our minds, and delivering our prostituted souls to the lupanar; you, who submit our barbarisms to your wise judgement, stick our solecisms with your arrows, staunch our malodorous chasms, castrate our

Silenes, clap our Noahs into breeches, emasculate our macrological discourses, patch up our ellipses, curb our tautologies, temper our acyrologies, excuse our escrologies, pardon our perissologies, forgive our cacophonies. I, again, conjure you all, all of you in general and you in particular, Poliinnio: halt that slanderous rage and that criminal hatred you feel towards the most noble female sex; do not ruin all that the world possesses of beauty, all that which heaven contemplates with countless eyes. Pull, pull yourselves together and recover your wits, by which you might see that your animosity is nothing but a professed madness and frenetic passion. Is there anyone more senseless and stupid than a man who doesn't see the light? Can there be a madness more miserable than becoming, on account of sex, the enemy of nature herself, like that barbarous king of Sarza, who, having learned from your kind, declared:

> Nature can make nothing perfect, since she is herself a woman.[22]

Consider somewhat the truth, lift your eyes to the tree of the knowledge of good and evil, and note the contradiction and opposition that exists between the one and the other; see what men are and what women are. You hold, on one hand, the body, masculine, to be your friend, and the soul, feminine, your enemy. On one hand, you have chaos, masculine, and on the other, organization, feminine. Here, sleep, masculine; there, wakefulness, feminine. On one side, forgetfulness, and on the other, memory. Here, hate, there, friendship; on this side, fear, on the other, serenity; on one hand, rigour and on the other, kindness; here, anger, there, calm. On one side, error, on the other, truth; here, imperfection, there, perfection; here, hell, there, happiness; on this side, the Poliinnio the pedant, on the other side, Poliinnia the Muse. In short, all the vices, imperfections and crimes are masculine, and all the virtues, merits and goodnesses are feminine. Hence, prudence, justice, strength, temperance, beauty, majesty and dignity, both in grammatical gender and in our imagination, as well as in our descriptions and paintings, are all feminine. But to leave aside these theoretical reasonings concerning grammar and nomenclature so appropriate to your argument, and to come to what is natural, real and practical, one example alone should serve to bridle your tongue and shut your mouth, yours and those of your many cohorts: imagine if someone should ask where you will

[22] Ariosto, *Orlando furioso,* XXVII, 120, slightly adapted by Bruno, who has, 'Natura non può far cosa perfetta, / poi che natura femina vien detta.' Ariosto writes, 'veggo che (natura) non può far cosa perfetta, / poi che Natura femina vien detta.'

find a man who surpasses, or is even equal to, this celestial Elizabeth, England's ruler. She is so highly endowed, elevated, favoured, protected and supported by the heavens that physical or verbal efforts to overthrow her are both vain. There is no one in the kingdom so worthy and so heroic among the nobility, nor anyone so gifted among those who wear the gown, or so wise among the counsellors. For corporal beauty, knowledge of vernacular and learned tongues, grasp of the arts and sciences, vision in governing, grandeur of such great and long-lasting authority and other natural and civic virtues, the Sophonisbas, Faustinas, Semiramises, Didos, Cleopatras and all the earlier queens that Italy, Greece, Egypt and other parts of Europe and Asia can boast are trivial compared to her. Her results and her successes, which the present age cherishes with honest wonderment, bear witness to this. While across Europe's back flow the wrathful Tiber, the threatening Po, the violent Rhine, the bloody Seine, the turbid Garonne, the frenzied Ebro, the furious Tagus, the tumultuous Meuse and the unquiet Danube, she, with her splendid vision, has been able, for more than five lustres, to calm the great Ocean, which, in its constant ebb and flow, calmly and gladly gathers the beloved Thames to its bosom, flowing on unchecked and fearless, gaily and confidently twisting between its verdant banks. So then, to start over again …

ARMESSO. Quiet there, Filoteo, quiet. Do not strain yourself adding water to our ocean and light to our sun. Quit showing yourself so abstract (not to mention worse) in your polemic against those absent Poliinnios. Instead, give us some examples from the dialogues you have here, so we do not idle away our hours today.

FILOTEO. Take them and read.

End of first dialogue

Second dialogue

Speakers: Dicsono Arelio, Teofilo, Gervasio, Poliinnio

DICSONO. Please, master Poliinnio, and you, Gervasio, do not keep interrupting our discussions.

POLIINNIO. *Fiat* [Agreed].

GERVASIO. Surely I cannot stay quiet if he, the *magister* [master], speaks.

DICSONO. You say then, Teofilo, that everything which is not a first principle and a first cause, has a principle and a cause?

TEOFILO. Without the slightest doubt or dispute.

DICSONO. Do you believe, then, that whoever knows the things caused and principled, may know the cause and principle?

TEOFILO. It is not easy concerning the proximate cause and principle; and it is extremely arduous, even by way of traces, when dealing with the first cause and first principle.

DICSONO. So how do you conceive that things which have both a first and proximate cause and principle can be truly known if, as far as the efficient cause is concerned (which is one of the causes that contribute to the authentic knowledge of things), they remain hidden?

TEOFILO. I confess that it is an easy thing to set out a demonstrative doctrine, but the demonstration itself is hard. It is very easy to organize the causes, modes and methods of doctrines, but our method-makers and analysts then apply their instruments, the principles of their methods and art of arts poorly.

GERVASIO. Like men who know how to forge fine swords, but not to wield them.

POLIINNIO. *Ferme* [Certainly].

GERVASIO. Would that one could firmly shut your eyes and keep you from ever opening them again![1]

TEOFILO. That is why I say that the natural philosopher is not required to produce all causes and all principles, but merely the physical ones, and among them, only those that are principal or pertinent. Therefore, although their dependence on the first cause and first principle attributes them to that cause or that principle, there is not such a necessary relation that, from the knowledge of one, we can infer a knowledge of the other, and that is why we do not require that they be discussed within a single system.

DICSONO. How is that?

TEOFILO. Because from the knowledge of all dependent things, we cannot infer any cognition of the first principle or of the first cause, other than by the less effectual method of vestiges; seeing that everything derives from its will or goodness, which is the principle of its operation, whence proceeds the universal effect. The same can be said of artistic products, insofar as whoever sees the statue does not behold the sculptor, and the man who sees the portrait of Helen does not see Apelles, but only the result of an operation deriving from the excellence of Apelles' talent. The representation is wholly the effect of accidents and circumstances of the

[1] The latin *ferme* is taken jokingly by Gervasio as a term related to the Italian *fermare*, 'to shut'.

substance of that man who, in terms of his absolute essence, is totally unknown.

DICSONO. So that to know the universe is to know nothing of the being or of the substance of the first principle, because it is like knowing the accidents of the accidents.

TEOFILO. Correct. But I would not want you to think that I mean there are accidents in God, or that he could be known through his accidents.

DICSONO. I do not ascribe to you such dull wit, and I know that it is one thing to say that all things not belonging to the nature of God are accidents, and another to say they are his accidents, and still another thing to say that they are like his accidents. This last is what I believe you are claiming for the effects of the divine operation: although they are the substance of things, or rather the natural substances themselves, they are nevertheless like accidents that are too remote to allow us to achieve cognitive apprehension of the divine, supernatural essence.

TEOFILO. Well put.

DICSONO. Of the divine substance, therefore, because it is both infinite and extremely remote from those effects which constitute the outer limit of the path of our discursive faculty, we can know nothing, except by means of vestiges, as the Platonists say, or of remote effects, as the Peripatetics have it, or by means of garments, as the Cabalists say, or of dorsal and back parts, as the Talmudists say, or of a mirror, shadow and enigma, as the Apocalyptics claim.

TEOFILO. But there is more: since we do not see that universe perfectly, of which the substance and principle are so hard to understand, we have far less basis for knowing the first cause and principle by means of its effects than we have of knowing Apelles through the statues he creates; for we can see the entire statue and examine it part by part, but not so the vast and infinite consequence of divine power. The resemblance, then, must be understood as not involving proportionality.

DICSONO. So it is and so I understand it.

TEOFILO. Therefore, we shall do well to abstain from discussing such a lofty subject.

DICSONO. I agree with that, because it suffices, morally and theologically, to know the first principle in so far as the heavenly gods have revealed it and the prophets have borne witness to it. Not only every law and every theology, but all reformed philosophies conclude that it is the token of a wroth and sacrilegious spirit to rush into demanding reasons and giving definitions of things above the sphere of our intelligence.

TEOFILO. Good. But these people do not deserve reproach, so much as those deserve the highest praise who strive towards the knowledge of this principle and this cause, to apprehend its grandeur as far as possible by inspecting, with the eyes of orderly consideration, those magnificent stars and luminous bodies which are so many inhabited worlds, great creatures and superlative divinities: those which seem to be, and are, innumerable worlds not very unlike that in which we find ourselves. Since it is impossible for them to have being in and of themselves, being composite and dissoluble (not that they are, therefore, deserving of dissolution, as was well expressed in the *Timaeus*), it is necessary that they have a principle and cause, and that, as consequence of the greatness of their being, living and acting, they manifest and proclaim in an infinite space and with innumerable voices the excellence and infinite majesty of their first cause and first principle. Leaving aside, then, as you say, that speculation, since it surpasses all sense and intellect, let us look into the principle or cause insofar as, as vestige, either it is nature itself, or it shines in the element and the bosom of nature. Question me, then, methodically, if you want me to answer likewise.

DICSONO. So I will. But first of all, since you frequently employ the terms 'cause' and 'principle', I would like to know whether you consider them synonymous.

TEOFILO. No.

DICSONO. But then what difference is there between the two?

TEOFILO. When we say that God is first principle and first cause, we mean one and the same thing, using different concepts, but when we speak of principles and causes in nature, we are talking of different things using different concepts. We say that God is first principle, in so far as all things come after him according to a definite order of anteriority and posteriority, in terms of either their nature, their duration or their merit. We speak of God as first cause, in so far as all things are distinct from him, as the effect from the efficient cause, and the thing produced from its producer. And these two definitions are different, because not everything which is prior and of higher value is the cause of what comes after it and is of lesser value, and because not every cause is prior and of higher value than that which is caused, as is clear to whoever ponders the matter carefully.

DICSONO. Then, tell me, what difference is there between cause and principle, as far as natural things are concerned?

TEOFILO. Although the terms are sometimes used interchangeably, nonetheless, speaking properly, not everything that is a principle is a cause. The point is the principle or origin of the line, but not its cause; the instant is the principle or origin of activity [but not the cause of the act]; the point of departure is the principle of movement, and not the cause of movement; the premises are the principles of an argument, but not its cause. 'Principle' is, thus, a more general term than 'cause'.

DICSONO. Then, to narrow these two terms within certain proper meanings, observing the procedure of those who express themselves most correctly, I think you take 'principle' to be that which intrinsically contributes to the constitution of things and remains in the effect, as they say of matter and form, which remain in the composite, or else the elements from which a thing is composed and into which a thing is resolved. You call 'cause' that which contributes to the production of things from outside, and which exists outside the composition, as is the case of the efficient cause, and of the end to which the thing produced is directed.

TEOFILO. Very good.

DICSONO. Now that we have answered the question of the difference between these things, I would like you to turn your attention first to the causes, and then to the principles. Regarding the causes, I would first like to know about the first efficient cause, then the formal, which you say is linked to the efficient, and lastly the final cause, understood as the mover of the efficient cause.

TEOFILO. The order of your proposition pleases me very much. As for the efficient cause, I say that the universal physical efficient cause is the universal intellect, which is the first and principal faculty of the world soul, which, in turn, is the universal form of it.

DICSONO. You seem to me to be not only in agreement with Empedocles' opinion, but still more certain, precise, explicit and even, as far as I can see by your statements, more profound. Thus, I would appreciate it if you would explain in greater detail just what you conceive this universal intellect to be.

TEOFILO. The universal intellect is the innermost, most real and most proper faculty or potential part of the world soul. It is that one and the same thing that fills everything, illuminates the universe and directs nature to produce her various species suitably. It is to the production of natural things what our intellect is to the production of the representations of things. The Pythagoreans call it the 'mover' and 'agitator of the universe'. As the poet has expressed:

> *totamque infusa per artus,*
> *mens agitat molem, et toto se corpore miscet.*[2]
>
> [pervading its members, mind stirs the whole mass and mingles with the whole body]

The Platonists call it 'world artificer'. They believe that it proceeds from the higher world, which is indeed one, to this sensible world, which is divided into many, and where, because of the separations of its parts, both harmony and discord reign. This intellect, infusing and instilling something of its own into matter, while itself remaining immobile and undisturbed, produces all things. The hermeticists say that it is 'most fecund in seeds' or yet that it is the 'seed sower', because it impregnates matter with all forms, which, according to their nature and manner of being, succeed in shaping, forming and weaving matter in ways that are so remarkable and numerous that they cannot be ascribed to chance, nor to any other principle incapable of differentiation and arrangement. Orpheus calls it 'the eye of the world', because it sees both the inside and outside of all natural things, in order that they may succeed in producing and maintaining themselves in their proper proportions, intrinsically as well as extrinsically. Empedocles calls it 'the differentiator', since it never tires of distinguishing the forms confused within nature's bosom, and of summoning the generation of one from the corruption of another. Plotinus says it is 'the father and progenitor', because it distributes seeds in nature's field and is the proximate dispenser of forms. As for us, we call it the 'internal artificer', because it shapes matter, forming it from inside like a seed or root shooting forth and unfolding the trunk, from within the trunk thrusting out the boughs, from inside the boughs the derived branches, and unfurling buds from within these. From therein it forms, fashions and weaves, as with nerves, the leaves, flowers and fruits, and it is from the inside that, at certain times, it calls back its sap from the leaves and the fruits to the twigs, from the twigs to the branches, from the branch to the trunk, from the trunk to the root. Similarly, in animals, it begins by deploying its work from the seed and from the centre of the heart, towards the outer members, and from these it finally gathers back towards the heart the faculties it had

[2] Virgil, *Aeneid*, VI, 726–7.

extended, as if it were twining up thread it had first unwound. Now, if we believe that intellect and reason are required to produce those works – dead works, so to speak – that we know how to fashion according to certain order and by imitation on the surface of matter, as when stripping and whittling a piece of wood we cause the shape of a horse to appear, how much superior must we hold that artistic intellect that, from the interior of the seminal matter, solders together the bones, extends the cartilage, hollows the arteries, airs the pores, interweaves the fibres, branches out the nerves and arranges the whole with such praiseworthy mystery? How much greater an artificer, I say, is he who is not limited to one part of matter, but works, continually present in the whole, on the whole! There are three kinds of intellect: the divine, which is everything, the mundane, of which we have just spoken, which makes everything, and the other, particular ones, which become everything, because a middle term is needed between the extremes, and this is the true efficient cause, not only extrinsic, but also intrinsic, of all natural things.

DICSONO. I would like to hear you distinguish between your conception of an efficient cause as extrinsic cause, and your idea of it as intrinsic cause.

TEOFILO. I call a cause extrinsic when, as efficient, it is not a part of the things composed and things produced; it is intrinsic in so far as it does not operate on the matter or outside it, but in the way we have just described. Thus, it is an extrinsic cause by its being, which is distinct from the substance and the essence of its effects, and because its being does not resemble that of things susceptible to generation and corruption, although it operates in those things; a cause is extrinsic, with regard to the action of its operation.

DICSONO. It seems to me you have spoken enough about the efficient cause. Now I would like to understand what that formal cause could be which you maintain is linked to the efficient cause: is it, perhaps, the ideal reason? For every agent which operates by means of a regulating intellect strives to produce its effects only by means of some intention, and this is impossible without the apprehension of something, which is none other than the form of the thing to be produced. Consequently, that intellect, which possesses the faculty of producing all species, and of sending them forth with such fine architecture from the potency of matter to act, must contain them all in advance, after the manner of forms, without which the agent could not proceed to manufacture, just as the sculptor cannot execute different statues without having a preconception of their different forms.

TEOFILO. You understand all that excellently. What I want, indeed, is that two sorts of form be considered: one is the cause which, even if not the efficient, allows the efficient to produce its effects; the other is the principle, called forth from matter by the efficient cause.

DICSONO. The aim, the final cause which is sought by the efficient, is the perfection of the universe, which consists of all forms having actual material existence; the intellect delights and takes such pleasure in pursuing this goal, that it never tires of calling forth from matter all sorts of forms, as Empedocles himself seems to maintain.

TEOFILO. Quite right, and I add that, just as this efficient is universal in the universe, but specific and particular in the universe's parts and members, so are also its form and its purpose.

DICSONO. But enough concerning causes. Let us come to principles.

TEOFILO. In order to get at the constitutive principles of things, I will first discuss form, since, in a way, it is identical to the efficient cause we have just defined: we said, in fact, that the intellect, which is a potency of the world soul, is the proximate efficient cause of all natural things.

DICSONO. But how can the same subject be principle and cause of natural things? How can it have the character of an intrinsic part, and not that of an extrinsic part?

TEOFILO. That is no contradiction, if we consider that the soul is in the body as the pilot is in the ship: since the pilot is part of the ship, he moves with it; yet, considering that he governs and moves it, he must not be included as a part, but as a distinct efficient cause. Likewise, the soul of the universe, in so far as it animates and informs it, is found to be an intrinsic and formal part of the universe, but in so far as it directs and governs the universe, it is not a part, and does not have the character of principle, but of a cause. Aristotle himself grants us this, since, though he denies that the soul has the same relation to the body as the pilot to the ship, he does not go so far, when he considers it with regard to its power to know and to understand, as to call it the act and form of the body, but he looks on it as an efficient cause separate in its being from matter. The intellect is something that comes from outside from the point of view of its substantiality, independent of the composite.

DICSONO. I approve of what you say, because if it is correct that the intellectual potency of our soul is separated from the body and has the character of an efficient cause, it is all the more true of the world soul. As Plotinus writes against the Gnostics, 'the world soul governs the universe more easily than the soul governs our body', since there is a great difference between their ways of governing. The former rules the world without being fettered to it, so that what it controls does not bind it, nor does it suffer through or with other things. It raises itself without impediment to higher things; giving life and perfection to bodies, it does not itself become infected with any imperfection: and that is why it is eternally united with the same subject. As for the latter, it is clear that its condition is completely different. Now, if, according to your principle, the perfections found in inferior natures must be attributed to, and recognized in, superior natures to a higher degree, we must agree, without the slightest doubt, with the distinction you have established. This assertion is valid not only for the world soul, but also for every star, since (as the aforementioned philosopher holds) they all have the power to contemplate God, the principles of all things and the distribution of the orders of the universe. He holds that this does not occur by means of memory, reasoning or reflection, for all their operations are eternal operations; no act can be new to them, and, in consequence, they do nothing which is inappropriate to the whole, nor anything which is not perfect or does not follow a definite and predetermined order, and all this completely without any act of deliberation. This is what Aristotle himself shows with the examples of the perfect writer or perfect lute player, when he denies that, under the pretext that nature does not reason or reflect, one can conclude that it operates without intellect or final intention: for great musicians and writers pay less attention to what they are doing than their less talented colleagues, who, because they reflect more, produce work that is less perfect and, what is worse, not free from error.

TEOFILO. You have understood. But let us look at things a bit more closely now. It seems to me that those who will not understand or affirm that the world and its parts are animated detract from the divine goodness and from the excellence of this great living being and simulacrum of the first principle; as if God were jealous of his image, as if the architect failed to love his own work – he of whom Plato[3] remarks that he appreciated his creation for its resemblance to himself, for the reflection of himself he sees in it. And, indeed, what could be presented to the eyes of the divinity which is more beautiful than this universe? And since the universe is composed of its parts, which of these parts should we hold to be more important than

[3] *Timaeus* 29E.

the formal principle? I leave for a better and more detailed analysis the examination of the thousand natural reasons that can be added to this topical or logical one.

DICSONO. I am not concerned to have you exert yourself over this point, for there is no philosopher enjoying some reputation, even among the Peripatetics, who does not hold that the world and its spheres are animated in some way. For now, I would like to understand how, in your opinion, this form comes to introduce itself into the matter of the universe.

TEOFILO. It joins itself to it in such a way that the nature of the body, which is not beautiful in itself, comes to participate as far as it can in beauty, for there is no beauty which does not consist of some species or form, and there is no form that is not produced by the soul.

DICSONO. I seem to be hearing something very novel. Are you claiming, perhaps, that not only the form of the universe, but also all the forms of natural things are souls?

TEOFILO. Yes.

DICSONO. But who will agree with you there?

TEOFILO. But who could reasonably refute it?

DICSONO. Common sense tells us that not everything is alive.

TEOFILO. The most common sense is not the truest sense.

DICSONO. I can readily believe that that last point is defensible. But it is not enough for one to be be able to defend a thing to render it true: we must be able to provide a proof.

TEOFILO. That is not difficult. Are there not philosophers who say that the world is animated?

DICSONO. Many of the leading ones do say so.

TEOFILO. Then, why do those same philosophers not declare that the world's parts are animated?

DICSONO. They, indeed, say it, but only of the principal parts, those which are the true parts of the world, for when they affirm that the soul is entire in the entire world, and entire in any of its parts, they are as reasonable as when they hold that the soul of living creatures we can perceive is wholly present throughout their bodies.

TEOFILO. So, which do you think are not true parts of the world?

DICSONO. Those which are not primary bodies, as the Peripatetics call them: the Earth, with the waters and other parts that, as you say, constitute the entire creature, along with the moon, the sun and other bodies. Besides these principal organisms, there are those that are not primary parts of the universe of which, it is said, some have a vegetative soul, others a sensitive soul, still others an intellective soul.

TEOFILO. But if the soul, present in the whole, is also in the parts, why do you not admit it in the parts of the parts?

DICSONO. I do, but only in the parts of parts of animate things.

TEOFILO. But what are these things that are not animated, or that are not parts of animated things?

DICSONO. Do you not think a few of them are right before our eyes? All lifeless things.

TEOFILO. And which things do not possess life, or at least the vital principle?

DICSONO. So, in short, you hold that there is nothing that does not possess a soul and that has no vital principle?

TEOFILO. Yes, exactly.

POLIINNIO. Then a dead body has a soul? So, my clogs, my slippers, my boots, my spurs, as well as my ring and my gauntlets are supposedly animated? My robe and my pallium are animated?

GERVASIO. Oh, yes, indeed, Master Poliinnio, why not? I do believe your robe and mantle are fully animated when they contain such an animal as you; the boots and spurs are animated when they cover the feet, the hat is when it covers the head, which is not deprived of a soul; the stable is animated also, when it shelters the horse, the mule or your lordship. Is that not what you mean, Teofilo? Do you not think I understand it better than the *dominus magister* [chief master]?

POLIINNIO. *Cuium pecus?* [Whose cattle?] Do we not find asses *etiam atque etiam* [several times] subtle? You have the nerve, you apirocal[4], you abecedarian, to compare yourself with an archididascalos[5] and rector of a minerval[6] school such as myself?

GERVASIO. *Pax vobis, domine magister, servus servorum et scabellum pedum tuorum.* [Peace be with you, lord master, I am your servant's servant and the footstool of your feet.]

POLIINNIO. *Maledicat te deus in secula seculorum.* [May God curse you, world without end.]

DICSONO. No fighting: allow us to settle these questions.

POLIINNIO. *Prosequator ergo sua dogmata Theophilus.* [Then let Teofilo continue to expound his theory.]

[4] Hellenism for 'ignoramus'. [5] Another hellenism formed from διδάσκαλος, 'teacher'. [6] Humorous latinism meaning 'literary'.

TEOFILO. So I will. I say, then, that the table is not animated as table, nor are the clothes as clothes, nor leather as leather, nor the glass as glass, but that, as natural things and composites, they have within them matter and form. All things, no matter how small and minuscule, have in them part of that spiritual substance which, if it finds a suitable subject, disposes itself to be plant, or to be animal, and receives the members of such or such a body, commonly qualified as animated, for in all things there is spirit, and there is not the least corpuscle that does not contain within itself some portion that may animate it.

POLIINNIO. *Ergo, quidquid est, animal est.* [Therefore, whatever is, is animal.]

TEOFILO. Not all things that have a soul are called animate.

DICSONO. Then, at least, all things have life?

TEOFILO. All things that have a soul are animated, in terms of substance, but their life is not recognizable to the Peripatetics, who define life too strictly and grossly, using the extrinsic and sensible act and operation, and not the substance.

DICSONO. You reveal a plausible way of supporting Anaxagoras' opinion that all things are in all things, for since the spirit, or soul, or the universal form is in all things, everything can be produced from everything.

TEOFILO. That is not only plausible but true, for that spirit is found in all things which, even if they are not living creatures, are animate. If not according to the perceptible presence of life and animation, then according to the principle, and a certain primary act of life and animation. I will go no further, since I wish to look later at the properties of many stones and gems which, broken, recut or set in irregular pieces, have certain virtues of altering the spirit or of engendering affections and passions in the soul, not only in the body. And we know that these effects do not, and could not, proceed from purely material qualities, but must be attributed to a symbolic principle of life and animation. Besides, we see the same phenomenon sensibly working in withered plants and roots which, purging and concentrating humours and altering their spirits, reveal unmistakable signs of life. Not to mention that necromancers, not without reason, hope to accomplish many things using the bones of the dead, believing that they retain, if not the very activity of life, at least some sort of vitality, which can be used to achieve extraordinary effects. Other occasions will give me the chance more fully to discuss thought, the spirit, the soul, the life which penetrates all, is in all, and moves all matter, fills its bosom, and dominates it rather than being dominated by it, given that the spiritual substance cannot be surpassed by the material substance, but, rather, contains it.

DICSONO. This seems to me to conform not only to the thought of Pythagoras, whose thesis the Poet states when he says:

> *Principio caelum ac terras camposque liquentis,*
> *lucentemque globum lunae Titaniaque astra*
> *spiritus intus alit, totamque infusa per artus*
> *mens agitat molem, totoque se corpore miscet;*[7]
>
> [First, the heaven and earth, and the watery plains,
> the shining orb of the moon and Titan's star,
> a spirit within sustains, and mind, pervading its members,
> sways the whole mass and mingles with its mighty frame.]

but also the thought of the Theologian, who says, 'the spirit pervades and fills the Earth, and what contains all things.'[8] Also, another, speaking perhaps of the relationships of form with matter and potency, says that act and form dominate.

TEOFILO. If, then, spirit, soul, life is found in all things and in varying degrees fills all matter, it can assuredly be deduced that it is the true act and true form of all things. The world soul, therefore, is the formal constitutive principle of the universe and all it contains. I say that if life is found in all things, the soul is necessarily the form of all things, that form presides everywhere over matter and governs the composites, determines the composition and cohesion of the parts. That is why it seems that such form is no less enduring than matter. I conceive this form in such a way that there is only one for all things. But according to the diversity of the dispositions of matter and the capacity of the material principles, both active and passive, it happens to produce different configurations and realize different potentialities, bringing forth sometimes non-sensitive life, sometimes sensitive but not intellective life, sometimes seeming to suppress or restrain all outside signs of life, because of the incapacity or some other characteristic of matter. Thus, changing site and state, this form cannot be annihilated, because spiritual substance is no less real than material. So only the external forms are changed, and even annihilated, because they are not things, but of things, and because they are not substances, but accidents and particularities of substances.

[7] Virgil, *Aeneid,* VI, 724–7.

[8] The author, supposed to be Solomon, of the Wisdom of Solomon, I, 7.

POLIINNIO. *Non entia sed entium.* [Not entities, but of entities.]

DICSONO. Certainly, if something of the substances were annihilated, the world would be emptied.

TEOFILO. Thus, we have an intrinsic formal principle, eternal and subsistent, incomparably superior to that imagined by the Sophists[9] who, ignoring the substance of things, treat only of the accidents, and arrive at positing corruptible substances from the fact that what they call essentially, fundamentally and principally substance is what results from composition, which is only an accident, not containing in itself either stability or truth, and reduced to nothing. They say that what is truly man is the result of composition, and that what is truly soul is no more than the perfection and act of a living body, or even something that is the result of a certain symmetry in its constitution and members. Hence, it is not surprising that they make so much, and are so greatly afraid, of death and dissolution, since they believe the loss of being is imminent. Nature cries out against such madness, assuring us that neither the body nor the soul need fear death, because both matter and form are absolutely unalterable principles:

> *O genus attonitum gelidae formidine mortis,*
> *quid Styga, quid tenebras et nomina vana timetis,*
> *materiam vatum falsique pericula mundi?*
> *Corpora sive rogus flamma seu tabe vetustas*
> *abstulerit, mala posse pati non ulla putetis:*
> *morte carent animae domibus habitantque receptae.*
> *Omnia mutantur, nihil interit.*

> [You people, dismayed by fear of icy death, why are you terrified by the Styx, by shadows and empty names, the stuff of poets' tales, by the dangers of a world that doesn't exist? Our bodies, whether destroyed by the flames of the funeral pyre, or by slow decay, do not feel any suffering. Our souls are immortal and are ever received into new homes, where they live and dwell, when they have left their previous abode. All things change, but nothing dies.][10]

DICSONO. I believe Solomon, esteemed the wisest among Hebrews, says something comparable: '*Quod est quod est? Ipsum quod fuit. Quid est quod fuit? Ipsum quod est. Nihil sub sole novum*' [What is that which is? That which was. What is the thing that was? That which is. There is nothing new under the sun].[11] And, thus, this form which you posit is not something

[9] For Bruno, the followers of Aristotle.

[10] Ovid, *Metamorphoses*, XV, 153–9 and 165, cited from memory.

[11] Ecclesiastes, 1, 9, cited from memory.

that exists only in matter and is fixed to it according to its being, and does not depend on the body or matter for its existence?

TEOFILO. Indeed. What is more, I leave open the possibility of a form existing without matter, though I firmly assert that no part of matter exists without form, except when it is considered logically, as it is by Aristotle, who never tires of dividing by reason what in nature and in truth is indivisible.

DICSONO. Do you not admit any other form than this eternal associate of matter?

TEOFILO. Yes, and a still more natural form than the material form, of which we shall treat later. For now, note this distinction of forms: first, there is a sort of form which informs, and which is extended and dependent. Since it informs everything, it is in everything; since it is extended, it communicates the perfection of the whole to the parts; since it is dependent and has no operation through itself, it communicates the operation of the whole to the parts; similarly it communicates the name and the being to them. Such is the material form, like that of fire: because every part of fire warms, is called fire, and is fire. Secondly, there is another sort of form, which informs and is dependent, but which is not extended. Since it perfects and activates the whole, it is in the whole and in each of its parts. Since it is not extended, the result is that it cannot attribute the action of the whole to the parts. Since it is dependent, it communicates the action of the whole to the parts. Such is the vegetative and sensitive soul, since no part of the animal is animal, yet each part nonetheless lives and feels. Thirdly, there is another kind of form, which actuates and makes perfect the whole but is not extended nor dependent as regards its operation. Since it perfects and actuates, it is in the whole, in its totality and in each of its parts. Since it is not extended, it does not attribute the perfection of the whole to the parts. Since it is not dependent, it does not communicate its action to them. Such is the soul, in so far as it can exercise intellectual power, and it is called intellective: it does not cause any part of man to be called man, or to be man, nor to be described as intelligent. Of these three kinds, the first is material, for it cannot be conceived, nor can it exist, without matter. The two other kinds (which, in fact, come together as one, according to their substance and being, and are distinguished in the fashion we have indicated above) express the formal principle, as distinct from the material principle.

DICSONO. I understand.

TEOFILO. Furthermore, I would like to point out that if, using common terms, we say there are five grades of form, namely, the elemental, the mixed, the vegetative, the sensitive and the intellective, we do not, however, understand form in the vulgar sense, for that distinction is valid from the point of view of the operations which appear in the subjects and proceed from them, but not from the point of view of the primordial and fundamental essence of that form and spiritual life that fills all things, which it does in different ways.

DICSONO. I understand. Inasmuch as this form that you posit as principle is a substantial form, it makes up a perfect species, is to itself its own genus, and is not, like the Peripatetic form, part of a species.

TEOFILO. Exactly.

DICSONO. The distinction of forms in matter is not a function of the accidental dispositions which depend on material form.

TEOFILO. Correct.

DICSONO. From which it follows that this separated form cannot be multiplied in the numerical sense, since all numerical multiplication depends on matter.

TEOFILO. Yes.

DICSONO. Further, it is invariable in itself, but varies because of the subjects and the diversity of matter. And such form, although it differentiates the part from the whole in the subject, is not, however, itself different in the part and in the whole, even if the definition we use for it differs according to whether it is considered as substantial by itself, or considered in so far as it is the act and perfection of some subject – and in that case, it is considered according to that specification and that individuation which it has assumed in this or that subject.

TEOFILO. Exactly.

DICSONO. You do not conceive of this form as being accidental, not as accidental form, not as mingled with matter or inherent in it, but as existing in it, associating itself with and assisting it.

TEOFILO. That is just what I say.

DICSONO. Moreover, this form is defined and determined by matter, since, on the one hand, possessing in itself the faculty of constituting the particulars of innumerable species, it happens to restrict itself in order to constitute an individual, and, on the other hand, the potency of indeterminate matter, which can receive any form whatsoever, finds itself limited to a single species. Thus, one is the cause of the definition and determination of the other.

TEOFILO. Very good.

DICSONO. So that, in some way, you allow Anaxagoras' notion, who qualifies with the word 'latent' the particular forms of nature, and you approve, in part, that of Plato, who has them derive from ideas, and, in part, that of Empedocles, who has them issue from the intelligence, and, in certain measure, that of Aristotle, who puts them down as emerging, so to speak, from the potency of matter?

TEOFILO. Yes, for as we have said, where there is form, there is, in a way, everything. Where there is soul, spirit, life, there is everything. The one who forms is the intellect, which acts through the ideal species; even if it does not bring forms out of matter, it does not then look for them outside of matter, since this spirit fills everything.

POLIINNIO. *Velim scire quomodo forma est anima mundi ubique tota* [I would like very much to know how the world soul is a form which is present everywhere in its totality], if it is indivisible. It must, then, be very large, even of infinite dimension, since you say the world is infinite.

GERVASIO. Here is good reason, indeed, for its being large. It is like what a preacher at Grandazzo in Sicily said of our Lord: to signify that he is present everywhere, he ordered a crucifix as large as the church, in the image of God the Father, who has the heavens for a canopy and the starry sky for a seat, and who possesses such long legs that they stretch down to earth, which he uses as a footstool. To this preacher came a certain peasant, saying, 'Reverend father, how many ells of cloth would it take to make his hose?' Another said that all Melazzo and Nicosia's chickpeas, haricots and broad beans would not suffice to fill his belly. Be careful, then, that this world soul is not cut out the same way.

TEOFILO. I would not know how to satisfy your perplexity, Gervasio, but I can that of Master Poliinnio. But for both of you, I will use a comparison, for I want you, too, to gather some fruit from our reasonings and discussions. In short, you must know, then, that the world soul and the divinity are not present entirely everywhere and through every part, in the same way as some material thing could be – since that is impossible for any body or spirit of any kind whatsoever – but are present in a manner that is not easy to explain, save in the following way. Please note that if we say the world soul and universal form are everywhere, we do not mean in a corporeal or dimensional sense, for they are not of that nature and cannot be found so in any part. They are everywhere present in their entirety in a spiritual way. To take an example (crude as it is), you might imagine a voice which is entire inside the whole room, and in every part of it: in effect, one hears it everywhere entirely there. Similarly, the words I am saying are entirely heard by all of you, and would still be if a thousand people were present. And if my voice could reach all the world, it would be everywhere entire. To you, then, I say, Master Poliinnio, that the soul is not indivisible in the manner of a point, but, in some way, in the manner of a voice. And to you, Gervasio, I answer that the divinity is not everywhere like the God of Grandazzo is in the whole of his chapel, because, though that God is present in the whole church, he is not everywhere wholly present, but has his head in one place, his feet elsewhere, his arms and his chest still somewhere else. On the contrary, the divinity is entire in any part whatsoever, just as my voice is heard entirely from all sides of the room.

POLIINNIO. *Percepi optime.* [I understood perfectly.]

GERVASIO. I have also understood your voice.

DICSONO. I believe it about the voice, but as for the argument, I think it has gone in one ear and out the other.

GERVASIO. I do not think it has even gone in, because the hour is late, and my stomach's clock has sounded.

POLIINNIO. *Hoc est, idest* [That is] he has his head *in patinis.* [on casseroles.]

End of second dialogue

Third dialogue

GERVASIO. It is already time and those people have not yet come. Since I have nothing else compelling to think about, I will amuse myself listening to their discussions, and maybe they can teach me some nice chess moves in the philosophy game besides. It is also a pleasant sport, with the whims that flit about in the bizarre brain of that pedant, Poliinnio. He presumes to be a judge of who speaks well, who discourses better, who commits philosophical incongruities and errors, but when his turn comes, and not knowing what to offer himself, he starts to spin out from the sleeve of his hollow pedantry a little salad of puny proverbs and phrases in Latin and Greek, which have nothing at all to do with what the others are saying, whence any blind man can see without much strain how mad he is, with his Latin, while the others are wise with their vulgar tongue. But, by my faith, here he is. By the way he moves along, it looks like he knows how to adopt a Latin pace even by the motion of his legs. Welcome, *dominus magister* [superior master].

POLIINNIO. That *magister* I do not care for. In our misguided and lawless age, it is attributed to any barber, rag picker or sow-gelder as often as to my peers. That is why we have the advice: *nolite vocari Rabi* [Be ye not called Rabbi][1]

GERVASIO. Then how do you wish me to call you? Do you fancy 'most reverend'?

POLIINNIO. *Illud est presbiterale et clericum* [That is for clerics and priests].

GERVASIO. Do you feel like 'most illustrious'?

POLIINNIO. *Cedant arma togae* [arms give way to the toga].[2] That title befits knights as well as the purple-clad.[3]

GERVASIO. And 'Caesarean majesty', hm?

POLIINNIO. *Quae Caesaris Caesari* [Give unto Caesar what is Caesar's].[4]

GERVASIO. Then take *domine* [Lord]!, take 'thunderer', *'divum Pater'* [father of the Gods]! To come back to us: why are you so late?

POLIINNIO. I think the others are held back by some business. As for myself, in order not to waste the day away without tracing a line,[5] I dedicated myself to contemplating that symbol of the globe commonly dubbed a world map.

GERVASIO. What on earth have you to do with a world map?

POLIINNIO. I am contemplating all the parts of the earth, climes, provinces and regions that I have traversed ideally in spirit, and many also on foot.

GERVASIO. I should like you to come back a little to yourself, as it seems to me that is what you need most but worry about the least.

POLIINNIO. *Absit verbo invidia* [Be it said without boasting][6]: in this way, I come to know myself much more effectively.

GERVASIO. And how, pray tell, is that?

POLIINNIO. From the scrutiny of the macrocosm, one can easily arrive (*necessaria deductione facta a simili* [having made the necessary deduction by similitude]) at the knowledge of the microcosm, the most minute parts of which correspond to parts of the former.

GERVASIO. So that, within you, we will find the moon, Mercury and other heavenly bodies, as well as France, Spain, Italy, England, Calcutta and other lands?

[1] From Matthew 23, 8. [2] Cicero, *fragm. De meo consulatu.* [3] Magistrates. [4] Matthew 22, 21. [5] Allusion to the Plinian adage: '*Nulla dies sine linea.*' [6] Livy IX, 19.

POLIINNIO. *Quidni? per quandam analogiam* [Why not? by a certain analogy].

GERVASIO. *Per quandam analogiam*, I believe you to be a great monarch, but if you were a woman, I would ask you if you have a place to lodge a baby, or stick one of those plants Diogenes speaks of.[7]

POLIINNIO. Ha, ha! *Quodammodo facete* [Quite prettily put]. But such questions are not befitting a sage and scholar.

GERVASIO. If I were a scholar, and if I considered myself wise, I would not come here to learn with you.

POLIINNIO. You, yes, but I do not come to learn, for *nunc meum est docere; mea quoque interest eos qui docere volunt iudicare* [my office now is to teach; my concern is also to pass judgement on those who wish to teach]. Hence, I come with another purpose than that which must bring you, whose role is that of apprentice, novice and disciple.

GERVASIO. For what purpose?

POLIINNIO. To judge, I said.

GERVASIO. Indeed, for one of your sort it is better to judge the sciences and doctrines than for others, since you are the only ones to whom the liberality of the stars and the munificence of fate have bequeathed the power to draw out the sap from words.

POLIINNIO. And consequently also from thoughts, which are bound to words.

GERVASIO. As soul to body.

POLIINNIO. Words that, rightly understood, give a thorough grasp of the sense: thus, from a knowledge of languages (in which I, more than anyone else in this city, am expert, and in which I count myself no less learned than any who run minerval schools) is derived the knowledge of all science.

GERVASIO. So, all who understand Italian will grasp the Nolan's philosophy?

POLIINNIO. Yes, but it also takes some exercise and judgement.

GERVASIO. A while ago, I thought that this exercise was the main thing, because someone who knows no Greek can nonetheless comprehend all Aristotle's meaning, and also pick out many of his errors. Similarly we see that the idolatry surrounding the authority of that philosopher (mainly regarding natural things) has been entirely abolished among all who grasp the notions of this other sect; a man who knows no Greek, nor Arabic, nor perhaps Latin, like Paracelsus, can have a better knowledge of

[7] Alludes to the obscene saying of Diogenes, '*planto hominem*'.

the properties of drugs and medicine than Galen, Avicenna and all those who communicate with the Roman tongue. Philosophies and laws are lost, not through a penury of word-interpreters, but through a scarcity of profound thinkers.

POLIINNIO. So, you number a man like me among the dull-witted multitude?

GERVASIO. God forbid. I know that, with study and knowledge (rare and remarkable things), you and your peers are thoroughly equipped to judge doctrines, after having sifted through the opinions of the people who champion them.

POLIINNIO. Since you are now speaking the pure truth, it is not so hard to persuade myself that you have some motive. If it is not too hard for you, pray take the trouble to set it out.

GERVASIO. I will say this (submitting myself throughout to your wise and prudent judgement): it is a common proverb that those who are not in a game follow it better than the ones playing. Similarly, those watching a play can better judge of the performance than the actors on the stage, and in the same way music can be better heard by someone not part of the orchestra or choir. It is the same with card games, chess, fencing and the like: and so, you other gentlemen pedants, excluded from all scientific and philosophical activity, not having nor ever having had anything to do with Aristotle, Plato and their kind, can better judge and condemn them with your grammatical matchlessness and natural presumption, than the Nolan, who finds himself on the same stage and in such familiarity and intimacy with them, having made out their most profound and innermost notions, that he fights them easily. I say that you, because you are outside every practice of gentlemen or extraordinary wits, can better judge them.

POLIINNIO. I have no idea how to answer this gross impudence point-blank. *Vox faucibus haesit* [The voice sticks in the throat][8].

GERVASIO. So, your sort possess that presumption lacking in those whose feet are deep into the question; therefore, I assure you it is with good title that you usurp the function of approving this, reproving that, glossing still the other, here drawing up a table of concordances, there an appendix.

POLIINNIO. This complete ignoramus wishes to infer from the fact that I am versed in letters that I am ignorant of philosophy!

GERVASIO. Most learned Poliinnio, sir, I must tell you that even if you knew all the languages there are, which our preachers number seventy-two ...

[8] Virgil, *Aeneid*, ii, 774.

POLIINNIO. *Cum dimidia* [and one half].

GERVASIO. ... not only would it not follow that you, sir, are capable of judging philosophers, but what is more, it would certainly follow that you cannot help being the biggest, most bumbling beast that exists in human form; besides, there is nothing to prevent anyone who has the least knowledge of any of these tongues, even a bastard one, from being the wisest and most learned man in the whole world. Consider how useful these two have been now: one, a French archpedant,[9] who has composed the *Studies in the Liberal Arts* and the *Animadversions Against Aristotle*, and another pedant scum, this one Italian, who has besmeared many an opuscule with his *Peripatetic Discussions*[10]. Everyone plainly sees that the first one very eloquently demonstrates his lack of intelligence, while the second shows that he has much in him of the beast and the ass, to put it bluntly. The first shows that he has understood Aristotle, at least, but badly. If he had understood him well, he might also have had the wit to wage honourable war with him, as the most judicious Telesio of Cosenza has.[11] Of the second, it is impossible to say whether he understood Aristotle either well or badly, but it can be claimed that he has read and re-read him, taken him apart, stitched him up again, and compared him *pro* [for] and *con* [against] with a thousand other Greek authors, going to the greatest lengths not only without any profit whatsoever, but *etiam* [even] to great loss. Whoever wants to see how far into insanity and presumptuous vanity a pedantic way of thinking can sink us has only to read this one book, before it disappears without a trace. But here come Teofilo and Dicsono.

POLIINNIO. *Adeste felices, domini* [you come at the right time, masters]: your arrival prevents my glowing anger from exploding into thundering judgements against the vain remarks issued by this sterile chatterer.

GERVASIO. And it blocks me from mocking the majesty of this most venerable owl.

DICSONO. All is well if tempers do not flare.

GERVASIO. What I say, I say in jest, because of the affection I feel for the honourable master.

POLIINNIO. *Ego quoque quod irascor, non serio irascor, quia Gervasium non odi* [That holds for me too. If I grow angry, my anger is not serious, for I do not hate Gervasio.]

[9] Peter Ramus (1515–72), author of *Scholae in Liberales* and *Aristoteliae Animadversiones.*

[10] Francesco Patrizi (1529–97). [11] Bernardino Telesio (1508–88).

DICSONO. Good. Let me take up my discussion with Teofilo.

TEOFILO. Thus, Democritus and the Epicureans, who claim that what is not body is nothing, maintain as a consequence that matter alone is the substance of things, and that it is also the divine nature, as an Arab named Avicebron has said in a book entitled *Fount of Life.* They also hold, together with the Cyrenics, the Cynics and the Stoics, that forms are nothing but certain accidental dispositions of matter. I, myself, was an enthusiastic partisan of this view for a long time, solely because it corresponds to nature's workings more than Aristotle's. But after much thought, and after having considered more elements, we find that we must recognize two kinds of substance in nature: namely, form and matter. For there must be an absolutely substantial act in which the active potency of everything is found, as well as a potency or substratum, in which an equal passive potency can be found: in the first, the power to make, in the second, the power to be made.

DICSONO. Anyone who reasons well will clearly see that it is impossible for the former continually to make everything, without there being something which can become everything. How can the world soul (I mean, all form), which is indivisible, act as shaper, without the substratum of dimensions or quantities, which is matter? And how can matter be shaped? Perhaps by itself? It seems we can say that matter is shaped by itself, if we want to consider as matter the universal formed body and call it 'matter', just as we would call a living thing with all its faculties 'matter', distinguishing it, not by the form, but only by the efficient cause.

TEOFILO. No one can keep you from using the term 'matter' as you wish, just as the same term covers different meanings in various schools. But I know that your way of considering it is only apt for a technician or physician strictly within his practice, for example that physician who reduced the universal body to mercury, salt and sulfur, a thesis that reveals the stupidity of his desire to be called philosopher more than some divine talent for medicine.[12] The aim of philosophy is not simply to arrive at the distinction of principles which is realized physically by the separation which results from the power of fire, but also to arrive at that distinction of principles to which no material agent can, since the soul, which is inseparable from sulphur, mercury and salt, is a formal principle; that principle is not susceptible to material qualities, but totally dominates matter and is not touched by the experiments of the alchemists, whose divisions are limited

[12] The reference is to Paracelsus.

to the three aforesaid elements, and who recognize another kind of soul, apart from this world soul, which we must define here.

DICSONO. Excellently said. And very satisfactory reasoning, for I see some people so lacking in judgement that they do not distinguish the causes of nature taken absolutely, according to the entire extension of their being, as philosophers do, and those taken according to a limited sense appropriate to their work. The first mode is excessive and vain for physicians as such, and the second is restricted and insufficient for philosophers as such.

TEOFILO. You have touched on that very point which earns Paracelsus praise. In discussing medical philosophy, he reproves Galen for having introduced philosophical medicine, and for having created such an annoying mixture and tangled web that, in the last analysis, he comes across as a very shallow physician and a very confused philosopher. But let that be said with some reserve, since I have not had the leisure to examine all parts of his work.

GERVASIO. Please, Teofilo, first do me the favour, since I am not so competent in philosophy, of making clear what you mean by the word 'matter', and what matter is in natural things.

TEOFILO. All who want to distinguish matter and consider it in itself, without form, resort to the analogy of art. So it is with the Pythagoreans, the Platonists and the Peripatetics. Take, for example, the art of carpentry: it has wood as substratum for all its forms and all its work, as iron is for the blacksmith and cloth for the tailor. All these arts produce various images, compositions and figures in their own particular material, none of which is natural or proper to that material. Nature is similar to art in that it needs material for its operations, since it is impossible for any agent who wishes to make something to create out of nothing, or to work on nothing. There is, then, a sort of substratum from which, with which, and in which nature effects her operations or her work, and which she endows with the manifold forms that result in such a great variety of species being presented to the eyes of reason. And just as wood does not possess, by itself, any artificial form, but may have them all as a result of the carpenter's activity, in a similar way the matter of which we speak, because of its nature, has no natural form by itself, but may take on all forms through the operation of the active agent which is the principle of nature. This natural matter is not perceptible, as is artificial matter, because nature's matter has absolutely no form, while the matter of art is something already formed by nature. Art can operate only on the surface of things already formed, like wood, iron, stone, wool and the like, but nature works, so to speak, from the centre of its substratum, or matter, which is totally formless. Furthermore, the substrata of art are many, and that of nature one, because the former, formed by nature in different ways, are diverse and various, while the latter, in no way formed, is undifferentiated throughout, since all difference or diversity proceeds from form.

GERVASIO. So that the things formed by nature serve as art's material, while a single, formless thing serves as nature's material.

TEOFILO. Yes.

GERVASIO. Can we know the substratum of nature, just as we can clearly see and know the substrata of the arts?

TEOFILO. Doubtless, but with different cognitive principles, for just as we do not know colours and sounds through the same senses, we cannot see the substrata of the arts and of nature with the same eye.

GERVASIO. You mean we see the first with the eyes of sense and the second with the eye of reason.

TEOFILO. Yes.

GERVASIO. Please elaborate.

TEOFILO. Gladly. The relation that the form of art has with its material is the same (allowing for proportions) as that of nature with its material. Just as in art, then, while the forms vary to infinity (if this were possible), under those forms there always persists one and the same matter – the form of the tree, for example, being followed by the form of the trunk, then of a board, then of a table, a stool, a chest, a comb and so on, while the wood remains the same – and it is no different in nature, where forms vary infinitely, one after the other, and the matter always remains the same.

GERVASIO. How can this analogy be confirmed?

TEOFILO. Do you not see that what was seed becomes stalk, what was stalk becomes an ear of wheat, what was an ear becomes bread, what was bread turns to chyle, from chyle to blood, from blood to seed, from seed to embryo, and then to man, corpse, earth, stone or something else, in succession, involving all natural forms?

GERVASIO. I see this easily.

TEOFILO. Then, there must exist one same thing which, in itself, is neither stone, nor earth, nor corpse, nor man, nor embryo, nor blood, nor anything else, but which, after having been blood, turns to an embryo by receiving the being of the embryo, and which, after having been an embryo, receives the being of man to become human, just as the matter formed by nature, which is the substratum of art, is a board and receives the being of board from what was a tree, and from the matter which was a board it receives the being of a door and is door.

GERVASIO. Now I understand it well. But it seems to me that this substratum of nature cannot be a body, nor have a definite quality. For, passing sometimes into such and such a form and natural being, sometimes into another form and being, it does not manifest itself corporally, like wood or stone, which always show through as they are, even if considered as material or substratum, no matter what the form.

TEOFILO. Well said.

GERVASIO. What shall I do, then, when I happen to be discussing this thought with some stubborn person who refuses to believe that there is only one matter underneath all nature's forms, just as there is only one beneath all the forms of each art? For we cannot deny what we see with our own eyes, but what we see solely through reason may be denied.

TEOFILO. Send him away or do not answer.

GERVASIO. But what if this stubborn individual demands evidence of this matter and is a respectable person, more liable to send me away than I am him, and takes my refusal to answer as an insult?

TEOFILO. What would you do if a blind demigod, worthy of every honour and respect, were so insistent, importunate and stubborn as to demand knowledge of and evidence for colours, or even for the external shapes of natural things? What if he asked, for example: What is the form of a tree? What is the form of mountains? of stars? Or again: What is the form of a statue, of a robe, or other artificial things, which are so plain to our eyes?

GERVASIO. I would tell him that if he had eyes, he would not ask for evidence of these things, since he could see them for himself, but since he is blind, it is impossible for others to show them to him.

TEOFILO. Likewise, you could say to your other people that if they had an intellect, they would not demand evidence of that natural matter, but could see it for themselves.

GERVASIO. Some would be humiliated by that answer, and others would see it as too cynical.

TEOFILO. Then, you can speak in less open fashion, as following: 'Most illustrious sir,' or 'Sacred Majesty, just as some things cannot be evident but for the hands and by the sense of touch, others only through hearing, others by taste, so this matter of natural things cannot be brought to light except through the intellect.'

GERVASIO. He may perhaps construe the shaft, which after all is neither very obscure nor very veiled, and answer, 'It is you who have no intellect: I have more than all your sort.'

TEOFILO. You will, then, credit him no more than you would a blind man who retorted that it is you who are blind, and that he sees much better than all those who believe they see like you.

DICSONO. You have said enough to demonstrate in more detail than I have ever heard what the word 'matter' means and what must be understood by 'matter' in natural things. In the same way Timaeus the Pythagorean[13] teaches us to find, through the metamorphosis of one element into another, the matter that is hidden and that can be known only in analogical terms. 'Where the form of the earth was', he says, 'there afterwards appeared the form of the water', and here we cannot say that one form receives the other, because a contrary thing does not accept or receive another. That is, the dry does not receive the wet, or rather the dryness does not receive the wetness, but there is a third thing from which the dryness is expelled and into which wetness is introduced, and this third thing is the substratum of both contraries, not being itself contrary to any. It follows that, since we cannot think of the earth as reduced to nothing, we must conjecture that something which was in the earth has subsisted and is found in the water. For the same reason, that same thing will subsist and will be found in the air, when the water is transmuted into air (under the effect of the heat which reduces it to fumes or vapour).

TEOFILO. From this we may conclude (in spite of our adversaries) that nothing is ever annihilated and loses its being, except for the external and material accidental form. That is why both the matter and the substantial form of any natural thing whatever (that is, its soul) can be neither destroyed nor annihilated, losing their being completely. Certainly this cannot be true of all the substantial forms of the Peripatetics and others like them, which consist of nothing but a certain complexion and a certain set of accidents; all that they are able to designate outside of their primary matter is nothing but accident, complexion, disposition of qualities, a principle of definition, quiddity. Hence, some cowled and subtle metaphysicians among them[14], wishing to excuse rather than accuse their idol Aristotle, have come up with humanity, bovinity, oliveness as specific substantial forms. This humanity – for example, Socratiety – this bovinity, this

[13] The pseudo-Timaeus of Locri, in his *De anima mundi et natura*, 94, A.
[14] The Franciscans of the school of Duns Scotus.

horseness, are individual substances. They have come up with all that in order to provide a substantial form which merits the name of substance, just as matter has the name of substance, and the being of substance. They have never derived any gain from this, for if you ask them, point by point, 'In what does the the substantial being of Socrates consist?', they will answer, 'In Socrateity'; if you then ask, 'What do you mean by Socrateity?', they will answer, 'The substantial form and proper matter of Socrates.' But let us leave aside this substance which is matter, and ask, 'What is the substance as form?' Some of them will reply, 'It is its soul'. Ask them, 'What is this soul?' If they say it is the entelechy and perfection of a body possessing potential life, remark that this is an accident. If they say it is a principle of life, sense, vegetation and intellect, remark that, although that principle is a substance if one considers it fundamentally, as we do, they present it as only an accident. For the fact of being a principle of such and such a thing does not express an absolute and substantial nature, but a nature that is accidental and relative to that which is principled: just as whoever says what I do or can do is not expressing my being and substance; that would be expressed by who says what I am, insofar as I am myself, considered absolutely. You see, then, how they consider this substantial form which is the soul: even if they have chanced to recognize it as substance, they have never, however, designated or considered it as such. You can make this conclusion out more plainly if you ask them in what consists the substantial form of an inanimate thing, for example, that of wood: the most subtle will imagine that it consists in woodness. Now take away that material common to iron, to wood, to stone, and ask, 'What substantial form of iron remains?' They will never point out anything but accidents. And these are among the principles of individuation, and provide particularity, because the material cannot be contained within the particular except through some form, and because this form is the constituent principle of some substance, they hold that it is substantial, but then they cannot show it physically except as something accidental. When they have finally done all they can, they are left with a substantial form which exists only logically and not in nature. Thus, a logical construction comes to be posited as the principle of natural things.

DICSONO. Aristotle does not realize this?

TEOFILO. I believe he fully realized it but could do nothing about it. This is why he says that the ultimate differences are unknown and cannot be expressed.

DICSONO. Then, he seems to me to have openly confessed his ignorance; therefore, I would be of the opinion that it is better to embrace those philosophical principles which, in this important question, do not plead ignorance, such as those of Pythagoras, Empedocles and your Nolan, whose opinions we touched on yesterday.

TEOFILO. This is what the Nolan holds: there is an intellect that gives being to everything, which the Pythagoreans and the *Timaeus* call the 'giver of forms'; a soul and a formal principle which becomes and informs everything, that they call 'fountain of forms'; there is matter, out of which everything is produced and formed, and which is called by everyone the 'receptacle of forms'.

DICSONO. This doctrine, from which it seems nothing is lacking, pleases me much. And, indeed, it is necessary that, just as we can posit a constant and eternal material principle, we similarly posit a formal principle. We see that all natural forms cease in matter, then appear again in matter; therefore, nothing, if not matter, seems in reality to be constant, firm, eternal and worthy to be considered as principle. Besides, forms do not exist without matter, in which they are generated and corrupted, and out of whose bosom they spring and into which they are taken back. Hence, matter, which always remains fecund and the same, must have the fundamental prerogative of being recognized as the only substantial principle; as that which is, and forever remains, and all the forms together are to be taken merely as varied dispositions of matter, which come and go, cease and renew themselves, so that none have value as principle. This is why we find philosophers who, having pondered thoroughly the essence of natural forms, such as one may see in Aristotle and his kind, have finally concluded that they are only accidents and particularities of matter, so that, according to them, it is to matter that we must accord the privilege of being act and perfection, and not to the things of which we can truly say that they are neither substance nor nature, but relative to the substance and nature – that is to say, in their opinion, matter, which for them is a necessary, eternal and divine principle, as it is to Avicebron, the Moor, who calls it 'God who is in everything'.

TEOFILO. Those who have not recognized any other form outside of accidental form have been led to this error, and this Moor, although he had accepted the substantial form from the Peripatetic doctrine in which he was nurtured, judged it corruptible and not merely susceptible to material mutations. Since he despised that which is produced and does not produce, is constituted and does not constitute, is remade but does not remake, he held it worthless compared to matter, which is stable, eternal, progenitor and mother. And this happens, inevitably, to those who do not know what we do.

DICSONO. The point has been very well examined. But it is time to turn from this digression back to our problem. We now know how to distinguish matter from form, as much from the accidental form (whatever it may be) as from the substantial form. We must still look into its nature and its reality. But first, I would like to know whether, in view of the great union that this world soul and universal form has with matter, one could not admit that other mode of philosophizing, belonging to those who do not separate the act from the essence of matter, and who understand matter as a divine thing, and not as something so pure and formless that it cannot form and clothe itself.

TEOFILO. It is not easy, because absolutely nothing operates on itself alone, and there is always some distinction between an agent and what is produced or that on which the action and operation work. For that reason, it is good to distinguish matter from soul in the body of nature, and in the soul to distinguish the idea of the species.[15] Hence we affirm that in this body there are three things: first, the universal intellect inherent in things; second, the soul that vivifies all; and third, the substratum. But we shall not refuse the name of philosopher to someone who follows his own bent and takes this formed body, or (as we prefer to call it) this rational animal, and then begins to take as first principle, in some sense, the constituents of this body, such as air, earth, fire, or even the ethereal region and the astral region, or spirit and body, or the void and the plenum (though not the void in Aristotle's sense), or yet another convenient way. Such a philosophy, it seems to me, does not deserve to be repudiated, especially when, no matter what basis is presupposed or what form of construction is contemplated, it helps to improve speculative science and knowledge of natural things, as was, indeed, done by many ancient philosophers. For it is a sign of an ambitious, presumptuous, envious and vain mind to wish to persuade others that there is only one way to investigate and to attain knowledge of nature, and it is the sign of a madman or man without reason to believe that this way lies within himself alone. So that, although we should always rather prefer, honour and practise the most resolute and constant, the most contemplative and dignified way, and the most lofty method of inquiry, we

[15] The universal intellect, mentioned in the following sentence.

should not fault that other method, which is not without fruit, though it does not belong to the same tree.

DICSONO. You approve, then, of the study of different philosophies?

TEOFILO. For those who possess an abundance of time and wit, I recommend it. For others, I approve the study of the best way, provided the gods allow them to guess which it is.

DICSONO. I am sure, however, that you do not approve of all philosophies, but only of the good or best.

TEOFILO. That is true. Likewise, among the different medical methods, I do not condemn the one that proceeds magically, applying roots, wearing stones or murmuring incantations, if the severity of the theologians will permit me to speak purely as a natural philosopher. I approve of what is done physically, carried out by means of apothecaries' prescriptions to flux or cause to run bile, blood, phlegm and melancholia. I accept that other method which proceeds alchemically, extracting the quintessences, and using fire to volatilize mercury, deposit salt, make sulphur grow luminous or extract oil from composites. But I do not wish to determine what method is the best among so many medical procedures. If the epileptic, to whom the physician and the alchemist have dedicated so much time in vain, chances to be cured by the magician, he will rightly endorse that medicine over the two other types. Keep to the same reasoning for other methods: none serves less well than another, if it attains the end it has set itself. In my case, I consider the doctor who cures me to be worth more than the others who hurt me or murder me.

GERVASIO. What is the reason for the great enmity between these schools of medicine?

TEOFILO. Greed, envy, ambition and ignorance. In the main, they hardly understand their own method of treatment, much less those of the other schools. Most of them try to get ahead by casting the others down and showing contempt for whatever they cannot acquire, being unable to elevate themselves to honour and profit thanks to their own merits. The best and truest among them is he who is not only physician, but also alchemist and astrologer. But, to return to our point, the best philosophy is that which brings about the perfection of the human intellect most easily and eminently, and most closely corresponds to the truth of nature. The best one renders us, as far as possible, co-operators with nature, whether by divination (I mean according to the natural order and the principles of change, not by animal instinct in the manner of beasts and those who resemble them, nor by the inspiration of good or bad demons, like the prophets, nor, finally, under the effect of melancholic enthusiasm, like poets or other contemplatives), or by instituting laws and reforming customs, by practising medicine, or even by becoming acquainted with and leading a blessed and more divine life. That is why no well-organized philosophy exists that does not contain some special quality not found in the others. I understand the same of medicine, which derives from principles that presuppose a fairly good philosophical outlook, as the function of the hand or foot presupposes that of the eye. Thus, it is said that there cannot be good medical principles where there is not a good point of departure in philosophy.

DICSONO. You please me greatly, and I praise you in equal measure, for just as you are not as vulgar as Aristotle, you are neither as pretentious nor offensive as he, devoting himself to belittling the opinions of all other philosophers as well as their manner of philosophizing.

TEOFILO. Of all the philosophers, I know none more reliant upon fancies and more remote from nature than he. Even if he says excellent things at times, it is recognized that they are not derived from his own principles, but are always propositions borrowed from other philosophers, such as those divine things we see in the books *On Generation, Meteors* and *On Animals and Plants.*

DICSONO. Coming back to the matter at hand, do you hold that one can give different definitions of matter, without error or contradiction?

TEOFILO. Yes, just as different senses may judge the same object and the same thing may reveal itself in diverse ways. In addition (as we have already mentioned), the same thing may be understood from different points of view. The Epicureans have said some good things, although they have not risen beyond the material quality. Heraclitus has shown us some excellent things, even though he has not gone beyond the soul. Anaxagoras manages to make progress in the study of nature, claiming to recognize, not only within nature but also outside and perhaps above it, an intellect which is the same as what Socrates, Plato, Trismegistus and our theologians call God. Thus, those who begin from an experimental analysis of simple elements (as they call them)[16] make as much progress in discovering the secrets of nature as those who start from a rational theory. And, among them, those who start from the study of physiological structure progress no less than those who begin from humours, and they, in turn, do

[16] I.e. the materialists: on one hand, Anaxagoras (who speaks of 'parts'), and on the other, the followers of Democritus and Epicurus (who speak of 'atoms').

no better than those who begin from the perceptible elements, or, more profoundly, from absolute elements, or from the one matter, which of all principles is the highest and most eminent. Sometimes, he who takes the longest way round does not make the best journey, especially if his purpose is not so much contemplation but action. As for the manner of doing philosophy, it will be no less advantageous to explicate forms as proceeding from something implicated than to distinguish them as from a chaos, or to distribute them, as from an ideal source, or have them pass into act as from some state of possibility, or draw them out as from a womb, or bring them out into the light as from a blind and gloomy abyss. For every foundation is good, if it is strong enough to support the edifice, and every seed is suitable, if the trees and the fruit are desirable.

DICSONO. To come now to our objective, please present us with your own detailed theory of this principle.

TEOFILO. Certainly, this principle, called matter, can be considered in two ways: first, as potency; second, as substratum. Regarded as potency, there is nothing in which it cannot be found in a certain way and in the appropriate sense; the Pythagoreans, the Platonists, the Stoics and others have placed it in the intelligible as well as in the sensible world. But we, who do not understand it exactly as they did, but in a more elevated and broader sense, speak of potency or possibility in the following way. Potency is commonly divided into active potency, through which its substratum can operate, and passive potency, through which it can exist, or receive, or have, or be the substratum of the efficient in some manner. Without taking active potency into consideration for the moment, I say that potency, in its passive sense (although it is not always passive), may be considered either relatively or absolutely. Thus, there is nothing to which we can attribute being without also attributing to it the possibility of being. And this passive potency corresponds so perfectly to active potency that one cannot exist in any way without the other, so that, if the power to make, produce and create has always existed, so, likewise, has the power to be made, produced and created, for one potency implies the other. I mean that in positing one we necessarily posit the other. Since this passive potency does not indicate any weakness in that to which it is attributed, but confirms, rather, its virtue and efficacy, and since the active potency and the passive potency are, in the end, one and the same thing, there is no philosopher or theologian who hesitates to attribute it to the first, supernatural principle. For the absolute possibility, through which the things that are in act can exist, does not come before the actuality, nor even after it. Furthermore, the power to be accompanies the being in act and does not come before it, for if what can exist made itself, it would exist before being made. Consider at present the prime and optimal principle, which is all it can be: if it could not be all, it would not be all; therefore, in it, act and potency are the same thing. This is not the case with other things, which, however much they are what they can be, could possibly not exist at all, or certainly could be something else, or be different from what they are, for nothing outside of the first principle is all that it can be. Man is what he can be, but not all that he can be. A stone is not all it can be, because it is not lime, nor dust, nor vase, nor grass. That which is all that it can be is a unity, which in its being comprises every being. Every other thing is not like that. That is why potency is not the same as act, since act is not absolute but limited. Moreover, the potency is always limited to a single act, because it never has more than one, specific and particular being. And even if it aspires to every form and every act, this is by means of certain dispositions, and following a certain succession of beings, one after another. Hence, every potency, every act which, in the principle, is (so to speak) enfolded, united and unique, is unfolded, dispersed and multiplied in other things. The universe, which is the great simulacrum, the great image and sole-begotten nature, is also all that it can be, through the very species and principal members, and by containing the totality of matter, to which nothing is added, nothing taken away, of complete and unified form. But it is also not all that it can be, because of its very differences, its particulars, its modes and its individuals. It is only a shadow of the first act and the first potency, and, in consequence, potency and act are not absolutely one and the same thing in it, since none of its parts is all that it can be. Furthermore, in the specific way that we have mentioned, the universe is all that it can be, in an unfolded, dispersed and distinct manner, while its first principle is all it can be in a unified and undifferentiated way, since all is there as a whole, an absolutely one and the same thing without difference or distinction.

DICSONO. What do you say of death, corruption, vices, defects, monsters? Do you think they also have a place in that which is all it can be and which is in act all it is in potency?

TEOFILO. These things are neither act nor potency, but defect and impotency found in unfolded things, because they are not all they can be and are compelled into becoming what they can be. Hence, unable to be many things at once, they lose one being in order to have another, and sometimes they confound one with another, sometimes becoming diminished, mutilated and maimed by the incompatibility of one being with another and by their occupation of the same matter. Getting back to our theme, the first absolute principle is greatness and magnitude, and it is a greatness and magnitude such that it is all it can be. It is not great with a greatness that can be superior or inferior, nor can it be divided, as can be every other greatness that is not all it can be. Hence, it is together maximum, minimum, infinite, indivisible by any measure. It is minimum, yet with nothing greater; maximum, yet with nothing smaller. It is beyond every equality, because it is everything it can be. And what I say of the greatness must also be understood for everything that can be said of it, because it is similarly the goodness which is every possible goodness, the beauty which is every possible beauty. There is no other beautiful thing that is all that it can be except for this one. The unity is that which is all, and can be all absolutely. Moreover, among natural things, we see none which is other than what it is in act; it is through the act that it is what it can be, from the fact that it possesses one kind of actuality. Nevertheless, even in its unique, specific being, no particular thing is all it can be. Take the sun: it is not all the sun can be, nor is it everywhere it can be. When it is east of the earth, it is not to the west, nor at midday, nor any other point. But if we want to show how God is sun, we will say (since he is all that he can be) that he is simultaneously in the east, west, noon, midnight and any other point whatsoever of the convexity of the earth. And so, if we wish to understand that our sun (either because of its own revolution or that of the earth) moves and changes position, because it cannot be found now at one point without being found potentially at all other points, and hence possesses a disposition to be at those points, if, therefore, the sun were all that it could be and possessed all that it was inclined to possess, it would be simultaneously everywhere and in all things; it would be so perfectly mobile and rapid that it would also be absolutely stable and immobile. Therefore, we find, in divine maxims, that the divinity is said to be eternally stable and absolutely rapid in its course from one end to the other.[17] For by immobile, we understand that which departs from and returns in the same instant to the eastern point, and which is not seen any less in the east than in the west or any other point of its circuit. That is why there is no basis on which to affirm that it goes and returns or has gone and returned from and towards such and such a point, rather than from and towards any other of the infinitely

[17] Book of Wisdom, 7, 24 and 7, 23.

numerous points. It will, therefore, be found entirely, and always, in the totality of the circle as well as in any of its parts; consequently, each individual point of the ecliptic contains the entire diameter of the sun. Thus, an indivisible is found to contain the divisible, and this is brought about not through any natural possibility, but through supernatural possibility – I mean, if one supposes this sun to be that which is in act all it can be. This absolute potency is not only what the sun can be, it is also what everything is and what everything can be. Potency of all potencies, act of all acts, life of all lives, soul of all souls, being of all beings; from whence the profound saying of the author of Revelation, 'He who is hath sent me to you; He who is speaks thus.'[18] And so, what is elsewhere contrary and opposed is one and the same in him, and every thing in him is the same. And you must reason regarding the differences of time and duration in the same manner as regarding the differences of actuality and possibility. He is, therefore, neither ancient, nor new, so that the author of Revelation describes him rightly as 'first and last'.[19]

DICSONO. This absolute act, which is identical with absolute potency, cannot be comprehended by the intellect, except by way of negatives: I mean, it cannot be grasped either in so far as it can be all, nor in so far as it is all things, for when the intellect wants to understand, it must try to form an intelligible species, and to assimilate and measure itself with that species. But this is an impossible task, for the intellect is never so great that it cannot be more so, while the absolute act, because it is immense on all sides and in all ways, cannot be greater. There is, then, no eye capable of approaching it or gaining access to such a sublime light and so profound an abyss.

TEOFILO. The coincidence of this act with absolute potency has been very plainly described by the divine spirit, when it says, *'Tenebrae non obscurabuntur a te. Nox sicut dies illuminabitur. Sicut tenebrae eius, ita et lumen eius'* [Yea, the darkness hideth not from thee, but the night shineth as the day: the darkness and the light are both alike to thee].[20] In conclusion, you see, then, how great is the excellence of the potency, and if you like to designate it the essence of matter, into which the vulgar philosophers have not penetrated, you may, without detracting from the divinity, treat it in a manner more lofty than has Plato in his *Republic* and his *Timaeus.* These works

[18] The author of the Revelation is the Holy Spirit speaking through scripture; the quote is adapted from *Exodus*, 3, 14.

[19] Frequent affirmation in the Bible, e.g. in *Isaiah* 41, 4.

[20] *Psalms* 139, 12.

have scandalized certain theologians because they have placed matter too highly. This has occurred either because these works were lacking in expression, or because the theologians, having been bred on Aristotle's opinions to consider matter solely in the sense of the substratum of natural things, have not understood them well. They do not see that, according to others, matter is understood as being common to the intelligible and sensible worlds (to use their terms, which give an equivocal meaning to matter based on an analogy). This is why opinions should be examined with great care before being condemned, and why it is necessary to distinguish terms as much as thoughts, for even if thinkers sometimes agree on a generic idea of matter, they go on to differ in their specific concepts. As for our argument, it is impossible that any theologian should be found (if we suppress the term 'matter', and however captious and malevolent his way of thinking) who would accuse me of impiety for what I say and think of the coincidence between potency and act, taking both terms in an absolute sense. Whence I would like to infer (in the measure allowed) that, in the simulacrum of that act and that potency, insofar as it is in specific act all that it can be in specific potency, the universe being all that it can be (let it be as it will in terms of the particular act and potency), there is a potency that is not separated from the act, a soul which is not separated from that which is animated – I mean, the simple, not the composite, so that the universe has a first principle taken as a unity, and no longer considered doubled into material principle and formal principle. This principle may be inferred by comparison with the aforesaid, which is absolute potency and act, so that it is neither difficult nor harmful to admit definitively that *as a substance, the whole is one*, as Parmenides, treated ignobly by Aristotle, perhaps conceived it.

DICSONO. You, therefore, hold that although, in descending along the ladder of nature, there are two substances, one spiritual and one material, both are eventually reduced to one being and one root.

TEOFILO. Yes, if you think that it can be tolerated by those who do not penetrate into the matter.

DICSONO. Very easily, provided that you do not raise yourself beyond the limits of nature.

TEOFILO. This has already been done. Since our conception or definition of the divinity differs from the common one, we have our personal definition, which is yet not so contrary or unfavorable to the other, and perhaps clearer and more explicit from the point of view of reason, which does not go beyond our subject matter, and from which I did not promise you I would abstain.

DICSONO. But enough has been said about the material principle, from the point of view of possibility or potency. For tomorrow, please prepare to go on to the consideration of the same principle from the point of view of being a substratum.

TEOFILO. I will.

GERVASIO. Goodbye.

POLIINNIO. *Bonis avibus* [May the omens be favourable for you].

End of third dialogue

Fourth dialogue

POLIINNIO. *Et os vulvae nunquam dicit: sufficit: id est, scilicet, videlicet, utpote, quod est dictu, materia* [And the womb never says, 'enough'. That is, namely, to wit, so to speak, that is to say, matter], which is designated by these terms, *recipiendis formis numquam expletur* [is never sated with receiving forms].[1] But since there is no one else in this Lyceum, *vel potius* [or rather] in this Anti-Lyceum, *solus (ita, inquam, solus, ut minime omnium solus) deambulabo, et ipse mecum confabulator* [I will stroll alone (in a solitude, I mean, in which I am anything but alone) conversing with myself]. Matter, then, is called by the prince of Peripatetics, of the great Macedonian,[2] the professor of transcendent genius, *non minus* [no less] than by the divine Plato and by others, *chaos*, or *hyle*, or *sylva* [chaos, material, abundant material], or mass, or potency, or aptitude, or *privationi admixtum* [mixed with privation], or *peccati causa* [cause of sin], or *ad maleficium ordinata* [disposed to evil], or *per se non ens* [not existing in itself], or *per se non scibile* [unknowable in itself], or *per analogiam ad formam cognoscibile* [knowable by analogy with form], or *tabula rasa* [a blank tablet], or *indepictum* [unmarked], or *subiectum* [subject], or *substratum*, or *substerniculum* [litter], or *campus* [field], or *infinitum*, or *indeterminatum*, or *prope nihil* [almost nothing], or *neque quid, neque quale, neque quantum; tandem* [neither what, nor which, nor how many; finally] after having taken aim with several comparisons between various disparate terms (in order to define its nature), it is called 'woman' *ab ipsis scopum ipsum attingentibus* [by those who hit right on target]; *tandem, inquam (ut una complectantur omnia*

[1] *Et ... sufficit:* from Proverbs 30, 16. [2] Aristotle, teacher of Alexander the Great.

vocula), a melius rem ipsam perpendentibus faemina dicitur [it has finished, I repeat, by being called woman (to gather everything into a single term) by those who have most effectively evaluated its very reality]. *Et mehercle* [And by Hercules], it is not without good reason that the senators of Pallas' realm[3] have judged it well to set matter and woman side by side, for they have been pushed to extremes of rage and frenzy by their dealings with the rigours of women – but just now an apt rhetorical flourish comes to mind. Women are a *chaos* of irrationality, a *hyle* [wood] of wickedness, a forest of ribaldry, a mass of uncleanliness, an inclination to every perdition (another rhetorical flourish here, called by some *complessio* [complexion])![4] Whence existed, in potency, *non solum remota* [not only remote], but *etiam propinqua* [also proximate], the destruction of Troy? In a woman. Who was the instrument of the destruction of Samson's strength? Of that hero, I mean, who became unvanquished conqueror of the Philistines with the famous ass's jawbone that he had found? A woman.[5] Who tamed, at Capua, the might and violence of that great captain and perpetual enemy of the Roman republic, Hannibal? A woman! *(Exclamatio!)* Tell me, O cytharist prophet, the reason for your weakness. '*Quia in peccatis concepit me mater mea*' ['in sin did my mother conceive me'].[6] O ancient forefather, first-made man, gardener of Paradise and cultivator of the Tree of Life, of what malice were you victim, to have been propelled with the entire human race into the bottomless gulf of perdition? *'Mulier quam dedisti mihi: ipsa, ipsa me decepit'* ['The woman that you gave me, it is she, she who deceived me'].[7] *Procul dubio* [Without doubt], form does not sin, and no form is a source of error unless it is joined to matter. That is why form, symbolized by the man, entering into intimate contact with matter, being composed or coupling with it, responds to the *natura naturans*[8] with these words, or rather this sentence: *'Mulier, quam dedisti mihi'*, *idest*, matter, which was given me as consort, *ipse me decepit; hoc est*, she is the cause of all my sins. Behold, behold, divine spirit, how the great practitioners of philosophy and the acute anatomists of nature's entrails, in order to show us nature plainly, have found no more appropriate way than to confront us with this analogy, which shows that matter is to the order of natural things what the female sex is to economical, political and civil order. Open, open your eyes and ...

[3] Philosophers.

[4] A *complexion* is a rhetorical figure in which the members of a period begin and end with the same term.

[5] Delilah, as told in Judges 16. [6] Psalms 51, 5. [7] See Genesis 3, 12–14.

[8] Scholastic formula designating nature as active power and producer.

Oh! I see that colossal idler, Gervasio, coming to snap the thread of my sinewy oration. I am afraid he has heard me, but what matter?

GERVASIO. *Salve, magister doctorum optime* [Good day, O great master of wise men]!

POLIINNIO. If you do not intend, *(tuo more)* [as is your custom], to jeer at me, *tu quoque, salve* [good day to you as well]!

GERVASIO. I would like to know what you were in the middle of mulling over alone.

POLIINNIO. As I was in my little interior temple of the Muses, *in eum, qui apud Aristotelem est, locum incidi* [I fell upon this passage in Aristotle], in the first book of the *Physics,* at the end, where the philosopher, wishing to elucidate what primary matter is, compares it to the female sex – that sex, I mean, which is intractable, frail, capricious, cowardly, feeble, vile, ignoble, base, despicable, slovenly, unworthy, deceitful, harmful, abusive, cold, misshapen, barren, vain, confused, senseless, treacherous, lazy, fetid, foul, ungrateful, truncated, mutilated, imperfect, unfinished, deficient, insolent, amputated, diminished, stale, vermin, tares, plague, sickness, death:

Messo tra noi da la natura e Dio
per una soma e per un grave fio.[9]

[By nature and by God among us sent
As a burden and heavy punishment.]

GERVASIO. I know you say this more to exercise yourself in the art of elocution and to show how ample and eloquent you are, than because you actually feel what you put into words. You humanists, who dub yourselves professors of the liberal arts, when you have gorged to the breaking point on notions, are in the habit of discharging them on poor women; just as when some other bile weighs on you, you pour it out onto the first student of yours who makes a mistake. But beware, you Orpheuses, of the furious wrath of the Thracian women.

POLIINNIO. I am Poliinnio, not Orpheus.

GERVASIO. Then, you do not really condemn women?

POLIINNIO. *Minime, minime quidem* [Not at all, indeed not at all]: I speak truly and mean nothing but what I say; for I do not (*sophistarum more*) [following the Sophists' custom], make a profession of demonstrating that white is black.

[9] Ariosto, *Orlando furioso,* XXVII, 119, quoted from memory. The actual lines run: 'Credo che t'abbia la Natura e Dio / prodotto, o scelerato sesso, al mondo / per una soma, per un grave fio'.

GERVASIO. Why do you dye your beard then?

POLIINNIO. *Ingenue loquor* [I speak sincerely], however, and I say that a man without a woman is like one of the intelligences; *qui non duxit uxorum* [he who has not taken a wife] is a hero, a demigod.

GERVASIO. He is also like an oyster, a mushroom, a truffle.

POLIINNIO. Whence the lyric poet has divinely declared:

Credite, Pisones, melius nil caelibe vita.[10]

[Believe me, O Pisones, there is no happier life than that of a celibate.]

And if you want to know the reason, listen to Secundus[11] the philosopher: 'Woman', he says, 'is an obstacle to calm, a continual havoc, daily warfare, a life-prison, a domestic storm, the shipwreck of man.' The man from Biscay[12] confirmed this when, angered by a terrible and furious storm at sea, at his wit's end he turned on the waves with a fierce and menacing look, saying, 'Ah, sea, sea, if only I could saddle you with a wife!' – to imply that woman is the tempest of tempests. That is why Protagoras, when asked why he had given his daughter to one of his enemies, replied that he could do him no greater harm than to furnish him with a wife. What is more, that good Frenchman won't call me to task when I say that when he received the order from Cicala,[13] the ship's master (with all those on board during a dangerous storm at sea), to throw their heaviest things overboard, he heaved his wife over at once.

GERVASIO. But you do not report the opposing cases of those who are very satisfied with their wives, among whom, under this very roof (to avoid going further), Monsieur Mauvissière. He has met with one who is not only endowed with uncommon physical beauty, which is veil and mantle of her soul, but who, furthermore, thanks to the triumvirate of penetrating judgement, heedful modesty, and very noble courtesy, holds her spouse's spirit bound with an indissoluble knot and has the ability to captivate all who meet her. And what will you say of her noble daughter, who has seen the light for scarcely a lustre and a year? You cannot tell whether his wife is from Italy, France or England, such is her linguistic talent; as for her touch with musical instruments, you cannot tell if she is a corporeal or incorporeal being; as regards her gifted manners, you wonder if she has really come of earth or dropped from the heavens. Everyone sees that

[10] Horace, *Epistles*, I, I, 88 and II, 3, 6. [11] Secundus of Athens, philosopher of the second century.

[12] Biscay merchants frequented England's ports at the time.

[13] Identified as a friend of Giovan Bruno, the author's father.

in her, just as the blood of her parents has mingled to produce her beautiful body, the virtues of their heroic spirits have fused to forge her extraordinary spirit.

POLIINNIO. *Rara avis* [Rare bird], that Marie de Bochetel. *Rara avis,* that Marie de Castelnau.[14]

GERVASIO. That *rare* you use for women can just as well be applied to men.

POLIINNIO. To get back to the point, a woman is but matter. If you do not know what a woman is because you do not know what matter is, study the Peripatetics a little; they will teach you what a woman is by teaching you about matter.

GERVASIO. I see that, with that Peripatetic brain of yours, you have learnt little or nothing from what Teofilo said yesterday about the essence and potency of matter.

POLIINNIO. Be that as it may. I hold to the point that one must condemn the appetite of both woman and matter, which is the cause of all evil, all affliction, defect, ruin and corruption. Do you not think that, if matter were satisfied with its present form, no alteration or affliction would hold sway over us, we would not die, we would be incorruptible and eternal?

GERVASIO. And what would you say if it were satisfied with the form it had fifty years ago? Would you be Poliinnio? If it had remained what it was forty years ago, would you be so adulterous (I mean, adult), so perfect and so learned? Thus, just as you are pleased that your other forms have given way to the current one, so it is nature's will, which orders the universe, that all forms yield to all others. Not to mention that it is much more dignified for that substance, which is our substance, to become everything by receiving all forms, than to remain fragmentary by holding onto only one. In that way, it shares a likeness with that which is all, in all.

POLIINNIO. It seems to be you are shedding your natural habits and beginning to be learned. Apply yourself, if you can, *a simili* [by similitude], to showing the dignity to be found in woman.

GERVASIO. That I will do easily. But here is Teofilo.

POLIINNIO. And Dicsono. Another time, then. *De iis hactenus* [Let us stop there].

TEOFILO. Have we not seen that the Peripatetics, like the Platonists, divide substance by the specific difference of corporeal and incorporeal? Just as these specific differences are reduced to the potency of a single

[14] Refers to the same woman, Marie de Castelnau, née Bochetel.

genus, so the forms must be of two kinds: some are transcendent, that is, higher than genus, and are called principles, such as 'entity', 'unity', 'one', 'thing', 'something', and their like; other forms belong to a given insofar as it is distinct from another genus, such as 'substantiality' and 'accidentality'. The forms of the first sort do not distinguish matter or make of matter here one thing, there another, but, as absolutely universal terms embracing corporeal as well as incorporeal substances, they signify the absolutely universal, absolutely common and undivided matter of both. Moreover, as Avicebron has said, 'Just as we identify the matter of the substantial form (matter which is part of the composite), before we identify the matter of the accidental forms (that is, the composite), what prevents us, before recognizing the matter that is contracted under corporeal forms, from recognizing a single potency, which is distinguishable through the form of corporeal nature and that of incorporeal nature, the one dissoluble, the other indissoluble?' Again, if everything that exists (beginning with the supreme and sovereign being) possesses a certain order and constitutes a hierarchy, a ladder where one climbs from the composite to the simple things, and from those to the most simple and absolute things, by means of proportional and copulative middle terms which participate in the nature of the one and the other extreme, yet possess their own, independent value, there is no order which does not involve a certain participation, nor participation which does not involve a certain union, nor union which does not involve a certain participation. It is therefore necessary that there be a single principle of subsistence for all existing things. Add to this the fact that reason, itself, cannot help presupposing, for anything which can be differentiated, something undifferentiated (I speak of things that exist, for I do not think the distinction between 'being' and 'non-being' is real, but merely verbal and nominal). This undifferentiated thing is a common nature to which the difference, the distinctive form, is joined. And surely one cannot deny that, since everything sensible presupposes a substratum of sensible matter, everything intelligible presupposes a substratum of intelligible matter. Therefore, something must exist which corresponds to the common nature of the one and the other substratum, for every essence is necessarily founded on some being, except for the first essence, which is identical with its being, since its potency is its act, and since it is all it can be, as we said yesterday. What is more, if matter is not a body (in the opinion of our adversaries, themselves), but by its nature precedes the corporeal being, why, then, would it be so inimical to the substances called incorporeal? Peripatetics are not lacking who hold that, just as in corporeal substances something formal and divine is found, so in divine substances something material should be found, so that the lower things should conform with the higher, and the order of the former should depend on that of the latter. As for the theologians, although some are nurtured on Aristotelian doctrine, if they will concede that they are more indebted to Scripture than to philosophy and natural reason, they should not be annoyed with me concerning this point. 'Do not worship me', said one of their angels to the patriarch Job, 'for I am your brother.'[15] Now, if the one who pronounces these words is an intellectual substance (for that it how they conceive it), and if he claims by his words that the man and he, himself, share in the reality of a substratum, whatever their formal differences may be, it follows that the oracle of these theologians testifies in favour of the philosophers.

DICSONO. I know you say that with reverence, since you know that it does not suit us to go begging in places outside our domain.

TEOFILO. You speak well and truly. But I did not bring in that reference to prove or confirm a point, but as far as possible to spare myself a scruple. I am just as afraid of appearing to be an enemy of theology as I am to be one.

DICSONO. Discerning theologians will always admit natural reasons, whatever course they may take, as long as those arguments do not go against divine authority.

TEOFILO. My arguments are and ever will be the same.

DICSONO. Good. Please go on.

TEOFILO. Plotinus, also, in his book on matter[16] says that 'if there is a multitude and a plurality of species in the intelligible world, there must be something common underlying the peculiarity and the difference of each. That which is common has the function of matter; that which is individual and which differentiates them has the function of form'. He adds that 'if this sensible world is an imitation of the intelligible one, the composition of one is an imitation of that of the other. Moreover, if the intelligible world lacked diversity, it would lack order, and if it lacked order, it would possess neither beauty nor ornament. All this is related to matter'. This is why the superior world should not be deemed totally indivisible, but in

[15] The angel does not speak to Jacob, but to Saint John in Apocalypse, XIX, 10, where, in any case, it is clear that the angel's speech is addressed not to Jacob, but to John himself.

[16] Enneads, II, 4, 4.

some ways divisible and differentiated – a division and a differentiation which are incomprehensible if there is not some underlying matter. And, although I claim that all this multiplicity comes together in a single indivisible being which is beyond any kind of dimension, I still assert that this being is the matter in which so many forms are united. Before it was conceived as being varied and multiform, it was conceived uniformly, and before being conceived as formed, it was conceived as unformed.

DICSONO. You have briefly set out many strong arguments enabling you to conclude that there is a single matter, a single potency, by which everything that exists does so in act. You also show that this applies equally to both corporeal and incorporeal substances, since the former have their being through their capacity to be, in the same way that the latter, through their capacity to be, have their being: all of which you have demonstrated by other strong arguments to those who ponder them deeply and fully grasp them. Even so, I would like you to spell out (if not for the sake of perfecting the doctrine, then at least for clarification) how there can be anything unformed and indeterminate in those most excellent beings which are the incorporeal things. How can they share in the same matter, without the advent of form and act resulting in bodies? How, when there is no mutation, generation, or corruption, can you say there is matter, when matter has never been posited for any other ends? How can we say that the intelligible nature is simple, and yet claim that matter and act are in it? I do not ask these questions on my own behalf, for whom the truth is clear; I ask, perhaps, for others who may be reluctant and difficult, like masters Poliinnio and Gervasio, for example.

POLIINNIO. *Cedo* [I concur].

GERVASIO. I approve, and thank you, Dicsono, for considering the needs of those who dare not ask, in keeping with the etiquette of transalpine meals, which forbids those who occupy the lesser seats at table to stick a finger outside the range of their own plates. There you must wait until a morsel is handed to you, and you cannot take a single bite without first having to pay for it with a 'thank you'.

TEOFILO. To resolve the whole question: just as a man, according to his specific human nature, is different from a lion, according to his particular nature, but both are indistinct and identical in their common animal nature, corporeal substance and other similar determinations, just so, according to its proper essence, the matter of corporeal things is different from that of incorporeal things. All that you say, then, concerning the fact of being a constitutive cause of corporeal nature, the fact of being the substratum of all sorts of transformations, and the fact of being a part of composites, agrees with matter in its proper essence. For the same matter (or, to put it more clearly), *the same* that can be made, or that can exist, is either made, and it exists through the dimensions and extension of the substratum and the qualities that have their existence in quantity – and this is called corporeal substance and presupposes a corporeal matter – or else it is made (supposing that its being has an inception) and is without those dimensions, extensions and qualities, and it is called incorporeal substance, and similarly presupposes the above mentioned matter. Thus, to an active potency, in the case of both corporeal and incorporeal things – that is, to both corporeal and incorporeal beings – there corresponds a passive potency, which is both corporeal and incorporeal, and a possibility of being which is both corporeal and incorporeal. If, then, we wish to speak of composition in the one nature as much as in the other, we must understand it in two different senses. We must also consider that, in eternal things, we speak of matter which is always under the same act, while in variable things, matter contains now one, now another act. With the former case, matter possesses, at once, always and together, all it can possess, and is all it can be; with the latter case, it has all it can possess and is all it can be, but at different times and according to a certain order of succession.

DICSONO. Some, though they admit matter in incorporeal things, understand it in a very different sense.

TEOFILO. However different their particular natures are, through which one thing descends to corporeal being and the other does not, and one thing receives sensible qualities and the other not, and however impossible it seems that there can be an essence common to, on the one hand, that matter which is incompatible with quantity and with the fact of being the substratum of qualities which have their existence in dimensions, and, on the other hand, that matter which is neither incompatible with the one nor with the other, nevertheless, they are one and the same thing, and the whole difference (as has been said many times) depends on the contraction of matter into corporeal being or incorporeal being. Similarly, in the animal being, all beings endowed with sense are one, but if we contract the genus to a particular species, the essence of a man is incompatible with that of a lion, and that of the lion with another animal. To this I add, if you please (since you might say that what is never found must be considered impossible, and unnatural rather than natural), that, primary matter never acquiring dimensions, one must consider corporeal matter as contrary to its nature, and that if this is so, it would not be likely that the two sorts of matter should have a common nature before one of them is conceived as being contracted to corporeal matter. I add, as I was saying, that we can just as well attribute to that first matter the necessity of having all dimensional acts, than (as you would have it) their impossibility. Since this matter is, in act, all that it can be, it has all measures and has all species of figures and dimensions. Because it has them all, it has none of them, since what is so many different things is necessarily none of them in particular. What is everything must exclude all particular being.

DICSONO. Do you claim, then, that matter is act? Do you also claim that matter in incorporeal things coincides with act?

TEOFILO. Yes, as the possibility to be coincides with being.

DICSONO. Then, it does not differ from form?

TEOFILO. It does not differ at all in the absolute potency and absolute act, and, because it is absolutely all, is therefore absolutely pure, simple, indivisible and unified. If it possessed definite dimensions, a definite being, a definite property and a definite individuality, it would not be absolute, nor would it be all.

DICSONO. Then, everything which comprises all the genuses is indivisible?

TEOFILO. Exactly, because the form which comprises all the qualities is itself none of them; that which comprises all figures does not itself possess any; that which possesses all sensible being is not, for that reason, accessible to the senses. That which possesses all natural being is highly indivisible; that which possesses all intellectual being is still more highly indivisible; that which possesses all that can be is the most highly indivisible of all.

DICSONO. You hold that there exists a ladder of the possibility to be, like the ladder of being? And you hold that material nature ascends along the one just as formal nature ascends along the other?

TEOFILO. That is true.

DICSONO. You give a lofty and profound definition of matter and potency.

TEOFILO. True again.

DICSONO. But this truth will not be grasped by everyone, for it is indeed hard to understand how it is possible to possess all the species of dimensions without having any, and to possess all formal being, and yet no form.

TEOFILO. Do you, yourself, understand how this can be?

DICSONO. I believe so, for I understand that, in order to be all things, the act cannot be any one thing.

POLIINNIO. *Non potest esse idem totum et aliquid; ego quoque illud capio* [The same thing cannot be, at the same time, the whole and some part of it. I, too, understand this].

TEOFILO. Then, you will be able to see how it follows that, if we wanted to posit having dimensions as the nature of matter, such a nature would not be incompatible with any kind of matter. But the only difference between the two matters is that one is freed from dimensions and the other is contracted to them. Being independent of dimensions, matter is above them all and comprehends them all; being contracted, it is comprehended by some dimensions and is under some of them.

DICSONO. You are right to say that matter, in itself, has no definite dimensions, and, therefore, must be understood as indivisible, receiving dimensions according to the nature of the form it receives. Its dimensions differ according to whether it is found under human form, under equine form, under that of the olive or under that of the myrtle tree. So, just as it has the faculty of receiving all those forms, before it exists under any of these forms, it has all of their dimensions in potency.

POLIINNIO. *Dicunt tamen propterea quod nullas habet dimensiones* [But that, they say, is because it possesses no dimensions].

DICSONO. And we say that *ideo habet nullas, ut omnes habeat* [it has no dimensions, so that it may have all of them].

GERVASIO. Why do you maintain that it includes, rather than excludes, all of them?

DICSONO. Because it does not receive dimensions as from without, but sends them out and brings them forth as from its womb.

TEOFILO. Well put. I might add that that is the way in which the Peripatetics habitually express themselves also, saying that the dimensional act and all natural forms emerge and derive from the potency of matter. Averroes understood this in part. Although an Arab, and not knowing Greek, he grasped more of the Peripatetic doctrine than any Greek we have read and he would have understood still more, had he not been so devoted to his idol, Aristotle. He says that matter, in its essence, comprises indeterminate dimensions. By this, he wishes to convey that they come to be determined – taking on now this figure and dimension, now others – according to the modification of natural forms. In this sense, we see that matter produces forms from itself, so to speak, and does not receive them as from outside. In a way, this is what Plotinus, prince of Plato's school, also understood. In establishing the difference between the matter of higher things and that of lower, he says that the first is everything at the same time and that, since it possesses all, there is nothing into which it changes, while the second, by means of a certain renovation at the level of parts, becomes everything, and becomes successively one thing after another – always, therefore, in a state of diversity, alteration and movement. In consequence, neither the one nor the other matter is ever formless, although each is formed differently; one in the instant of eternity, the other in the instant of time; one in simultaneity, the other in succession; one by way of enfolding, the other by way of unfolding; one as a unity, the other as multiplicity; one as being all and each thing, the other individually and thing after thing.

DICSONO. So, you wish to infer that, not only according to your principles, but also according to those of other philosophical methods, matter is not *prope nihil* [almost nothing], pure potency, bare, without act, without virtue and perfection.

TEOFILO. Exactly. I say that it is deprived of forms and without them, not in the way ice lacks warmth or the abyss is without light, but as a pregnant woman lacks the offspring which she produces and expels forth from herself, and as the earth is without light at night in our hemisphere, which it can reacquire by its turning.

DICSONO. So, even in these inferior things, act coincides in the end – if not entirely, at least to a great extent – with potency.

TEOFILO. I leave you to decide.

DICSONO. And what would happen if, finally, this potency from below became one with that from above?

TEOFILO. Judge for yourself. You can henceforth rise to the concept, I do not say of the supreme and most excellent principle, which has been excluded from our inquiry, but to the concept of the world soul, insofar as it is the act of everything and the potency of everything, and insofar as it is present in its entirety in everything – whence it follows that (even if there exist innumerable individuals) all things are one, and the knowledge of that unity is the object and term of all philosophies and all meditation on natural things – leaving in its domain the highest speculation of all, that which, surpassing nature, is impossible and vain for the unbeliever.

DICSONO. That is true, for one ascends there guided by supernatural and not natural light.

TEOFILO. It is that which is lacking in those who deem that everything is a body, simple like the ether, or composite like the stars and astral things – and who do not look for the divinity outside of the infinite world and the infinity of things, but inside that world and those things.

DICSONO. It is only on that point, it seems to me, that the faithful theologian differs from the truthful philosopher.

TEOFILO. I agree. I think you have understood what I mean.

DICSONO. Very clearly, I believe. And so, I infer from your remarks that, even if we do not let matter go beyond the level of natural things and keep to the common definition that the more vulgar philosophy gives of it, we will find that matter retains a greater excellence than is recognized in it by that philosophy. For, in the end, it does not attribute any other status to it except that of being a substratum of forms and a potency which is receptive to natural forms – without name, definition or determination because it is without any actuality. This point seemed difficult to certain monks[17] who, wishing to excuse rather than to accuse this doctrine, claimed that matter possesses only entitative act – that is, different from that which is simply without being and which has no reality in nature, as, for example, some chimera or imaginary thing. For this matter, in the end, has being – which is enough for it – similar to that which, without mode or dignity, depends on actuality and is nothing. But you could insist on asking Aristotle: Why do you claim, O prince of the Peripatetics, that matter is nothing, from the fact of its having no act, rather than saying that it is all, from the fact that it possesses all acts, or possesses them confusedly and confoundedly, as you prefer? Is it not you who, always speaking of the new being of the forms in matter, or of the generation of things, says that forms proceed from and emerge from inside matter? You have never been heard to say that forms proceeded – through the action of the efficient cause – from outside matter, saying rather that the efficient cause makes them emerge from within. I shall not mention that you also make an internal principle of the efficient cause of those things, to which you give the common name 'nature', and not an external principle as is the case with artificial things. In that case, it seems to me we should say that when matter receives a form from outside, it does not possess in itself any form or act. It also seems to me that when one says it sends all forms forth from its womb, we must declare that it possesses them all. Is it not you who, if not obliged by reason, at least compelled by normal usage, defines matter by saying that

[17] Followers of Duns Scotus.

it is 'that thing from which each natural species is produced', never saying that it is 'that in which things are made' – as we would say if acts did not come out of it and if, consequently, it did not possess them?

POLIINNIO. *Certe consuevit dicere Aristoteles cum suis potius formas educi de potentia materiae quam in illam induci, emergere potius ex ipsa quam in ipsam ingeri* [Certainly, Aristotle and his followers usually say that forms come out from matter, rather than that they are introduced into it, that they emerge from it rather than being absorbed into it], but I would say that Aristotle preferred to call 'act' the unfolding of form rather than its enfolding.

DICSONO. And I say that the expressed, sensible and unfolded being does not constitute the fundamental essence of actuality, but is a consequence and effect of it. In the same way, the principle being of wood and the essence of its actuality do not consist in being a bed, but in its being a substance so constituted that it can be a bed, a bench, a beam, an idol and anything else formed out of wood. Not to mention that all natural things are more genuinely produced from natural matter than artificial things are from artificial matter, for art generates forms from matter either by subtraction, as when it forms a statue from stone, or by addition, as when a house is formed by joining stone to stone and wood and earth. But nature produces everything out of its own matter by means of separation, parturition and effluxion, as the Pythagoreans thought, as Anaxagoras and Democritus understood and the sages of Babylon confirmed. Moses, himself, also subscribes to their opinion when, describing the generation of the things ordered by the universal efficient cause, he speaks thus: 'Let the earth bring forth its animals, let the waters bring forth living creatures.'[18] It is as if he had said: Let matter bring them forth. For, as he says, water is the material principle of things – which explains why he also says that the efficient intellect (which he calls spirit) 'brooded on the waters':[19] that is, he gave the waters a procreative power and produced from them the natural species, which he says afterwards are waters in substance. Thus, speaking of the separation of lower and higher bodies, he says, 'the spirit separated the waters from the waters', and deduces from this that dry earth appeared in their midst. Everyone claims, then, that things come from matter by way of separation, and not by means of addition and reception. Therefore, rather than saying that matter is empty and excludes forms, we should say that it contains forms and includes them. This matter which

[18] Paraphrase of Genesis 1, 20 and 24.

[19] Genesis 1, 2: Bruno translates *ferebatur* as *covava*, 'brooded'.

unfolds what it possesses enfolded must, therefore, be called a divine and excellent parent, generator and mother of natural things – indeed, nature entire in substance. Is that not what you mean, Teofilo?

TEOFILO. Certainly.

DICSONO. I am also very surprised that our Peripatetics have not worked out their art analogy further. Among the many materials that it recognizes and adopts, art considers that which is least subject to corruption and most durable and most versatile as best and most valuable. So, it deems gold more noble than wood, stone and iron, because it is less subject to corruption, and because everything that can be made of wood or stone, and many other things besides, can also be made of gold, producing things of much greater value by reason of their beauty, resistance, suppleness and nobility. What, then, must we say of the matter of which man, gold and all natural things are made? Must it not be held more worthy than the material of art, and must we not attribute a higher actuality to it? Why, O Aristotle, will you not admit that what is the foundation and the basis of actuality – I mean, of that which is in act – and which you declare to exist forever and endure eternally, why will you not admit that it is more in act than your forms and your entelechies, which come and go? So that if you also wanted to seek the permanence of the formal principle …

POLIINNIO. *Quia principia oportet semper manere* [Because principles should be permanent].

DICSONO. … without resorting to the fantastical ideas of Plato, since you are so hostile to them, you will be forced and obliged to say, either that the permanent actuality is found in the efficient cause – but this you cannot do, since you say that this efficient cause is what draws out and extracts the forms from the potency of matter – or that their permanent actuality is found in the bosom of matter. And, in fact, that is what you will be forced to say, because all the forms which appear as it were on the surface of matter – those that were as much as those that are or will be – and which you call individual forms in act, are not themselves the principle, but are principled things. (I think, in fact, that the particular form is found on the surface of matter, in the same way as the accident is at the surface of the composite substance. Whence it follows that the actuality of the expressed form must be recognized as inferior to that of matter, just as the actuality of the accidental form is recognized as inferior to that of the composite.)

TEOFILO. Indeed, Aristotle concludes lamely by declaring, in concert with all the ancient philosophers, that principles must always be permanent; later on, if we seek further in his doctrine for the place where the perpetual seat of natural form which floats on the back of matter might be, we will not find it either in the fixed stars – since the particular forms which we see do not descend from on high – nor in the ideal signs, separate from matter – for if these are not monsters, they are assuredly worse than monsters, being chimeras and pointless fantasies. What then? Forms are in the bosom of matter. And what then? Matter is the source of actuality. Do you want me to carry on and make you see all the absurdities into which Aristotle gets himself? He says that matter exists in potency, but ask him: When will it be in act? Together with a great crowd he will respond: When it possesses form. But insist and ask: When that occurs, what commences to exist? They will answer, despite themselves: The composite, not matter, since the latter is always identical to itself, never renews itself, never changes. The same goes for artificial things: when one makes a statue of wood, we do not say that the wood begins to exist, for it is no more nor less wood than before. In fact, that which receives being and actuality is the new product, the composite, I mean the statue. How can you grant potency, then, to something that will never be in act nor possess act? For it follows from this that matter is not that which is in potency of being or that can be, for it is always identical and immutable, and is that upon which and in which change takes place, rather than that which changes. What is altered, augmented, diminished, moved in location, corrupted, is always (as you Peripatetics, yourselves, say) the composite, never matter. Then, why do you say that matter is now in potency, now in act? No one surely could doubt that matter, whether it receives forms or sends them forth from itself, does not receive a greater or lesser actuality in terms of its essence or its substance; so that there is no reason to say that it exists in potency. For potency concerns what is in continual movement in relation to matter, and not matter itself, which is not only eternally at rest, but the very cause of that state of eternal rest. For if form, in keeping with its fundamental, specific being possesses, not only logically – in the concept and in reason – but also physically in nature, a simple, invariable essence, then form must exist in the perpetual potency of matter, which is a potency not distinct from act, as I have several times explained in my various discussions concerning potency.

POLIINNIO. *Queso* [I beg you], spare a word for the appetite of matter, so that Gervasio and I can resolve a little dispute between us.

GERVASIO. Yes, please, Teofilo, for this person has given me a pain in the head with his comparison between matter and woman. He says that women are no more content with males than matter is with forms, and so forth.

TEOFILO. Seeing that matter does not receive anything from form, why do you think it desires it? If (as we have said) it brings forms out of its bosom and so possesses them in itself, how can you claim that it desires them? It does not desire those forms which daily change on its back, for every ordered thing desires that from which it receives perfection. And what can a corruptible thing bring to an eternal one? What can an imperfect thing, as is the form of sensible things, which is always in movement, give to another so perfect that, if well pondered, is understood to be a divine being in things, as perhaps David of Dinant meant, who was so poorly understood by those who reported his opinion?[20] Matter does not desire form in order to be preserved by it, because a corruptible thing does not preserve an eternal one. Moreover, since matter clearly preserves form, form must desire matter in order to perpetuate itself, and not the other way around. For when form is separated from matter it ceases to exist, as is not the case with matter, which has all it had before the coming of form and which can have other forms as well. Not to mention that when we speak of the cause of corruption, we do not say that the form flees from matter or that it leaves matter, but that matter throws off one form to assume another. There is as little reason to say that matter desires form as that it hates it (I mean those forms that are generated and corrupted, because it cannot desire the source of forms, which it has within itself, because nothing desires what it possesses). By the same line of reasoning, according to which it is said to desire what it sometimes receives or produces, it can also be said to abhor whatever it throws off or rejects. In fact, it detests more fervidly than it desires, for it eternally throws off that individual form after retaining it a very short while. If you will remember this, that matter rejects as many forms as it assumes, you must agree with me when I say that it loathes form, just as I can allow your statements concerning desire.

GERVASIO. Here lie, then, in ruins not only Poliinnio's castles, but also others'.

POLIINNIO. *Parcius ista viris* [Do not boast too much].

DICSONO. We have learned enough for today. Until tomorrow.

20 David of Dinant, author of *De tomis idest de divisionibus.* 'Thou who reported his opinion' probably refers to Aquinas, *Summa contra Gentiles* (I, XVII), where David of Dinant is said to have identified God with matter.

TEOFILO. Then, goodbye.

End of the fourth dialogue

Fifth Dialogue

TEOFILO. The universe is, therefore, one, infinite and immobile. I say that the absolute possibility is one, that the act is one; the form, or soul, is one, the matter, or body, is one, the thing is one, being is one. The maximum, and the optimum, is one: it cannot be comprehended and is therefore indeterminable and not limitable, and hence infinite and limitless, and consequently immobile. It has no local movement since there is nothing outside of it to which it can be moved, given that it is the whole. It does not engender itself because there is no other being that it could anticipate or desire, since it possesses all being. It is not corrupted because there is no other thing into which it could change itself, given that it is everything. It cannot diminish or grow because it is an infinity to or from which nothing can be added or subtracted, since the infinite has no measurable parts. It is not alterable in terms of disposition, since it possesses no outside to which it might be subject and by which it might be affected. Moreover, since it comprehends all contraries in its being in unity and harmony, and since it can have no propensity for another and new being, or even for one manner of being and then for another, it cannot be subject to change according to any quality whatsoever, nor can it admit any contrary or different thing that can alter it, because in it everything is concordant. It is not matter, because it is not configured or configurable, nor it is limited or limitable. It is not form, because it neither informs nor figures anything else, given that it is all, that it is maximum, that it is one, that it is universal. It is neither measurable nor a measure. It does not contain itself, since it is not greater than itself. It is not contained, since it is not less than itself. It is not equal to itself, because it is not one thing and another, but one and the same thing. Being one and the same, it does not have distinct beings; because it does not have distinct beings, it has no distinct parts; because it has no distinct parts, it is not composite. It is limit such that it is not limit, form such that it is not form, matter such that it is not matter, soul such that it is not soul: for it is all indifferently, and hence is one; the universe is one.

Indeed, in this one, the height is no greater than the length or depth, so that it is called a sphere by analogy, although it is not a sphere. Length, breadth and depth in the sphere are identical, because they have the same limit, but in the universe, length, breadth and depth are identical because they are all equally without limit and infinite. If they have no half, quarter or other fraction, if there are no fractions at all, then there is no measurable part, nor, strictly, any part that differs from the whole. For, if you wish to speak of part of the infinite, you are obliged to call that infinite as well; if it is infinite, it coincides in one and the same being with the whole: therefore, the universe is one, infinite, indivisible. And if in the infinite you cannot find any difference as of part from whole, nor any difference as of one part from another, the infinite is undoubtedly one. There is no smaller part and greater part within the infinite's comprehension, for any part, however large, comes no nearer the proportion of the infinite than does any other, however small. In infinite duration, an hour is no different from a day, a day from a year, a year from a century, a century from an instant, because neither moments nor hours exist any more than do centuries, and because none is more commensurable with eternity than another. Similarly, in the immensity, the palm is not different from the stadium[1] , nor the stadium from the parasang[2] , because the parasang is no nearer the immensity's proportions than is the stadium. Hence, there are no more infinite hours than there are infinite centuries, nor infinite palms in greater number than infinite parasangs. You come no nearer to commensurability, likeness, union and identity with the infinite by being a man than by being an ant, or by being a star than by being a man, for you get no nearer to that infinite being by being the sun or the moon than by being a man, or an ant. This is because, in the infinite, there is no difference between those things – and what I say of them applies just as well to all other existent particular things.

Now, if, in the infinite, all these particular things are not differentiated, are not divided into species, it necessarily follows that they have no number: the universe, therefore, is one and immobile. Because it comprises everything, does not take on one being after another, and suffers no change neither by nor in itself, it is, consequently, all that it can be, and in it (as I said the other day), act does not differ from potency. If potency does not differ from act, it is necessary that, in the infinite, the point, the line, the surface and the body do not differ. For there, the line is surface because, by moving, it may become surface, and there the surface moves and becomes body, insofar as it may move and become, by its flow, a body. In the infinite, therefore, the point necessarily does not differ from the body, for, from its

[1] See note 3 on p. 10 above. [2] See note 3 on p. 10 above.

status as point, it becomes a line; from its status as line, it becomes a surface; from its status as surface, it becomes a body. So the point, because it possesses the potency to become a body, does not differ from the status of a body, where the potency and the act are one and the same thing.

The undivided does not differ, therefore, from the divided, nor does the absolutely simple differ from the infinite, nor does the centre differ from the circumference. Since the infinite is all that it can be, it is immobile; since in it everything is indifferent, it is one; and since it possesses all the greatness and perfection that can possibly be possessed, beyond all limit, it is the maximum and supreme immensity. If the point does not differ from the body, nor the centre from the circumference, nor the finite from the infinite, nor the maximum from the minimum, we may certainly affirm that the universe is entirely centre, or that the centre of the universe is everywhere, and the circumference nowhere insofar as it is different from the centre; or else that the circumference is everywhere, but the centre is nowhere insofar as it differs from the circumference. Here, then, is how it is not impossible, but rather necessary, that the optimum, the maximum, the incomprehensible is everything, is everywhere, is in everything, for, being simple and indivisible, it can be everything, be everywhere and be in everything. Thus, not for nothing is it said that Jove fills all things, inhabits all parts of the universe, is the centre of everything which has being: one in all, and that through which all is one, and is that which, being all things and comprehending all being in itself, causes everything to be in everything.

But you will say to me, 'Then why do things change? Why does particular matter turn itself into other forms?' My answer is that mutation is not striving for another being, but for another mode of being. And this is the difference between the universe and the things of the universe: for the universe contains all being and all modes of being, while each thing of the universe possesses all being but not all modes of being. Each thing cannot possess, in act, all particularities and accidents, because many forms are incompatible within the same subject, either because they are contrary or because they belong to different species – for example, there cannot be the same individual substance under the accidents of a horse and a human being, or under the dimensions of a plant or an animal. What is more, the universe comprehends all being totally, for nothing can exist outside of or beyond infinite being, because there is no outside or beyond infinity. By contrast, each of the things of the universe comprehends all being, but not totally, because outside each one of them there exists an infinity of other things. You must conceive, therefore, that everything is in everything, but not totally or under all modes in each thing. Understand, therefore, that each single thing is one, but not in the same way.

We are, therefore, correct in affirming that being – the substance, the essence – is one, and since that one is infinite and limitless, both with respect to duration and substance, as it is in terms of greatness and vigour, it does not have the nature of either a principle or of what is principled;[3] for each thing, coinciding in unity and identity (that is to say, in the same being), comes to have an absolute value and not a relative one. In the infinite and immobile one, which is substance and being, if there is multiplicity, the number which is a mode and multiformity of being by which it comes to denominate things as things, does not, thereby, cause being to be more than one, but to be multi-modal, multiform and multi-figured. So, following closely the reasoning of natural philosophers and leaving the logicians to their fantasies, we discover that everything that causes difference and number is pure accident, pure figure, pure complexion. Every production, of whatever kind, is an alteration, while the substance always remains the same, since there is only one substance, as there is but one divine, immortal being. Pythagoras, who did not fear death but saw it as a transformation, reached this conclusion. Those philosophers who commonly go by the name of physical philosophers[4] also were able to understand this. They said that nothing, in terms of substance, is begotten or is corrupted – unless we understand by this the process of change. Solomon inferred this as well, saying, 'there is nothing new under the sun, but what is, has already been'.[5] You see, then, how the universe is in all things and all things are in the universe, we in it and it in us: thus, everything coincides in perfect unity. See, then, how our spirit should not be afflicted, how there is nothing that should frighten us: for that unity is stable in its oneness and so remains forever. It is eternal, while every aspect, every face, every other thing is vanity and nothingness – indeed, outside this one there is nothing. Those philosophers who have discovered this unity have found their beloved Wisdom. For wisdom, truth and unity are indeed the same thing, though not everyone has understood this, since some have adopted the manner of speaking, but not the manner of understanding of the truly wise. Aristotle, among others, did not discover the

[3] I.e. there is no distinction between an active and passive component within infinite substance.
[4] Ionic philosophers. [5] Ecclesiastes, 1, 10.

one, nor being, nor the true, because he did not recognize being as one. Although he could have adopted the meaning of being which is common to substance and accident, and further, distinguished his categories according to as many genera and species as there are specific differences, nonetheless he perceived truth badly, not going deeply enough into the knowledge of this unity and of this indistinction of the eternal nature and eternal being. With his harmful explanations and his irresponsible arguments, this arid sophist perverted the sense of the ancients and hampered the truth, less, perhaps, out of intellectual weakness, than out of jealousy and ambition.

DICSONO. So that this world, this being, this truth, this universe, this infinity, this immensity is found entire in each of its parts: it is the *ubique* [everywhere] itself. Thus, everything in the universe, in relation to the universe, exists everywhere according to its capacity, whatever its relation might be with other particular bodies; for it is above, below, right, left and so on, in keeping with all local differences, since, in the totality of the infinite, there are all these differences and none of them. Whatever thing we take in the universe, it has in itself that which is entire everywhere, and hence comprehends, in its own way, the entire world soul (although, as we have said, it does not comprehend it totally), and that world soul is entire in every part of the universe. This is why, even if the act is one and constitutes a single being, wherever it may be found, we must not think that there is, in the world, a plurality of substance and of that which is truly being.

Following on this, I know that you take as manifest that each of these innumerable worlds, which we see in the universe, is not found there so much as if in a containing site, nor as in an interval or a space, but is found there as in a place that comprehends it, a conserver, mover and efficient, which itself is comprised in its entirety in each of these worlds, as the soul is found in its entirety in each of the parts of that world. For that reason, although a particular world moves towards or around another, as the earth moves to and around the sun, nonetheless, with respect to the universe, nothing moves to or around it, but only within it.

You, further, hold that, just as the soul (to take up the common way of speaking once more) pervades that great mass to which it gives being, remaining altogether indivisible, so that it is altogether present in the whole and any of its parts, so the essence of the universe is one both in the infinite and in anything taken as a member of the universe; so that, substantially, the whole and each of its parts are but one. In your opinion, Parmenides was, therefore, right to say that the universe is one, infinite and immobile (although it is not entirely clear what he intended, his words having been reported by a commentator who is not particularly reliable).[6]

You say that all the differences seen in bodies, from the point of view of formation, constitution, figures, colours and other individual or common characteristics, are nothing but the diverse aspects of the same substance: fleeting, mobile and corruptible aspects of an immobile, persistent and eternal being in which all forms, figures and members exist, though indistinctly and (so to speak) conglomerated – exactly as in the seed, where the arm is not distinct from the hand, nor the bust from the head, or the nerve from the bone, and where differentiation and separation do not produce another or a new substance, but bring into act and accomplish certain qualities, differences, accidents and dispositions related to that substance. And what is said of the relation between the seed and the members of animals may also be said of food in relation to chyle, to blood, to phlegm, to flesh and seed. This goes for any other thing which precedes the alimentary state, or other state. It also goes for all things, from the lowest level of nature to the highest, climbing from the physical totality philosophers know to the archetype in which theologians believe, if you like, until we reach a single original and universal substance, the same for all, which we call being, the basis of all species and all different forms. Similarly in the art of carpentry there is a single substance of wood which is subject to all dimensions and shapes, which are not themselves wood but are of wood, in the wood or involving the wood. That is why everything that makes for the diversity of genera, species, differences, properties, all that which consists in generation, corruption, alteration and change, is not being, is not essence, but condition and circumstance of being or essence, which is one, infinite, substratum, matter, life, soul, truth and goodness.

You say, then, that, since being is indivisible and absolutely simple, because it is infinite, and is act in its fullness in the whole and in every part of it (in the same way we speak of parts *in* the infinite, but not of parts *of* the infinite), we cannot think in any way that the earth is a part of being, nor that the sun is part of substance, since the latter is indivisible. But it is quite reasonable to speak of the substance of the part, or better still, of the substance in the part; just as it is not reasonable to say that a part of the soul is found in the arm or another part in the head, but it is legitimate to say that the soul is in the part that is the head, and that the substance is substance of the part – or in the part – that is the arm. For to be portion, part,

[6] Aristotle. See the end of the third dialogue, where Aristotle's opinion of Parmenides is mentioned.

member, the whole, equal to, larger or smaller, like this or like that, relative to this or to that, identical to or different from, etc., respond to other concepts which do not express an absolute, and hence cannot designate the substance, the one or being, but, in terms of modes, determinations and forms, exist through the substance, in the one, and relative to being. Thus, just as it is commonly said that quantity, quality, relation, action, passion and other kinds of accidents are relative to one and the same substance, in the same way one could say that the one and supreme being, in which act does not differ from potency, can be all absolutely and is everything that it can be. It is in a complicative manner one, immense, infinite and comprehensive of all being, and in an explicative manner, it is present in sensible bodies and in the potency and the act that we see distinguished in them. That is why you hold that what is generated and generates (be it a question of an equivocal agent or of a univocal agent, as is commonly said in philosophy), as well as that of which the generation is made, are always of the same substance. Your ears will, therefore, not be jarred by the thesis of Heraclitus, which declares all things to be but one – the one which, thanks to its mutability, contains all things in itself. And since all forms are in it, it follows that all definitions accord with it, so that all contrary propositions are true. And what creates multiplicity in things is not being, is not the thing, but what appears, what is offered to the senses and lies on the surface of things.

TEOFILO. Exactly. But I would like you to read further in this most important science and of this solid foundation of the truths and the secrets of nature. First, therefore, I would like you to note that nature descends to the production of things, and intellect ascends to the knowledge of them, by one and the same ladder. Both ways proceed from unity to unity, passing through a multitude of middle terms. Not to mention that the philosophical method of the Peripatetics and of many Platonists is to have the multitude of things as middle term, preceded by the pure act, at one extremity, and the pure potency, at the other; similar to other philosophers who affirm metaphorically that the darkness and the light come together in the constitution of innumerable degrees of forms, images, figures and colours. But beside all these philosophers, who take into consideration two principles and two princes, others rise up who, impatient with and hostile towards polyarchy, make the two principles coincide into one, which is at the same time abyss and darkness, clarity and light, profound and impenetrable obscurity, and supernal and inaccessible light.

Secondly, consider that the intellect, wishing to liberate and detach itself from the images to which it is bound, not only resorts to mathematical and symbolic figures or analogies drawn from them in order to comprehend the being and the substance of things, but also ascribes the multiplicity and diversity of species to one and the same root. Thus, Pythagoras, who posited numbers as the exclusive principles of things, understood unity to be the basis and substance of all of them. Thus, Plato and other philosophers who made species to consist of figures conceived of the point as substance and universal genus, inasmuch as it is the common stock and root of all figures. And perhaps surfaces and figures are what Plato meant ultimately by his 'great', and the point and the atom are what he meant by his 'small', two principles of specification of things which refer, then, to one, as everything that is divided refers to the undivided. Therefore, those who say that the one is the substantial principle mean that substances are like numbers, and others who think of the substantial principle as a point mean that the substances of things are like figures, but all agree in positing an indivisible principle. However, Pythagoras' method is better and purer than Plato's, because unity is the cause and the reason for individuality and the point, and it is a principle which is more absolute and appropriate to universal being.

GERVASIO. Why has Plato, who came after him, not done as well or better than Pythagoras?

TEOFILO. Because he preferred to speak less well, in a manner less adequate and less appropriate, and to be acclaimed as a master, than to say something better, in a better manner and be reputed a disciple. I mean that the goal of his philosophy was more his personal glory than the truth; seeing that, as I cannot doubt, he knew very well that his manner was more appropriate to corporeal things or things considered corporeally, while that of Pythagoras was no less suitable and adequate for corporeal things than it was for those things which reason, imagination, intellect, and both intelligible and sensible nature can forge. As everyone will acknowledge, Plato was not ignorant of the fact that unity and numbers are essential in order to justify and explain points and figures, but that these latter are not essential for justifying and examining unity and numbers, as dimensional and corporeal substance depends on the incorporeal and the indivisible. Furthermore, he knew that unity and number are independent from points and figures, because numbers may be explained without reference to measure, but measure is not independent from numbers, because the understanding of measure cannot be found without an understanding of numbers. That is why arithmetical analogy and proportion are better suited than geometry to guide us, by means of multiplicity, in the contemplation and apprehension of that indivisible principle which, because it is the unique and radical substance of all things, cannot possess a distinct and limited name, or any term that has a positive rather than privative meaning. Therefore, it has been called by some 'point', by others 'unity', and by still others 'infinity', and so on, with various like terms.

Add to what has been said, that when the intellect wishes to grasp the essence of something, it proceeds by simplifying as much as possible: I mean that it shuns composition and multiplicity, rejecting accidents, which are corruptible, as well as dimensions, signs and figures, and turns to what lies beneath these things. Just as a lengthy, long-winded oration cannot be understood but by reducing it to a simple conceit. By so doing, the intellect clearly demonstrates how the substance of things consists of unity, which it looks for either in reality, or by analogy. The man who could reduce to a single proposition all the propositions disseminated in Euclid's principles would be the most consummate and perfect geometrician; likewise, the most perfect logician would be he who reduced all propositions in logic to one. Herein lies the level of intelligence, because inferior intellects cannot understand multiplicity except through many species, analogies and forms, superior intellects do better with less, and the very best do perfectly with very little. The premier intelligence embraces everything in a single, absolutely perfect idea, and the divine mind and the absolute unity, with no species, is that which understands and that which is understood simultaneously. So that, to ascend to perfect knowledge, we proceed by grouping and restricting the many, just as unity, descending to the production of things, proceeds by unfolding into many. The descent moves from a single being to an infinity of individuals and innumerable species; the ascent moves from the latter to the former.

Therefore, to conclude this second consideration, I say that when we aspire and strive towards the principle and substance of things, we progress towards indivisibility, and that we must never believe we have arrived at the first being and the universal substance until we have come to this indivisible one in which all is comprised. Meanwhile, let us not be led into believing we can understand of the substance and essence more than what we can understand concerning indivisibility. Peripatetics and Platonists gather the infinity of individuals into a simple concept, which is their species; they gather countless species under determined genera, which Archytas first declared to be ten in number; they gather the determined genera into one being, a single thing: but this thing, this being, is understood by them as a name, a term, as a logical concept, and finally a vain thing. For then, when they treat of the physical, they no longer recognize a single principle of reality and of being for all that is, as they have recognized a common concept and name for all that which is expressible and intelligible. All this is due to their intellectual weakness.

Thirdly, you must know that substance and being are distinct from and independent of quantity, so that number and measure are not substance, but relative to substance; not being, but relative to being. We must define substance, therefore, as essentially without number and without measure and, consequently, as one and undivided in all particular things – which, themselves, owe their particularity to number, that is, to things relative to substance. Thus, whoever apprehends Poliinnio as Poliinnio does not apprehend a particular substance, but apprehends substance in the particular and in the differences which characterize it and which, by these differences, comes to place this man under a species in number and multiplicity.[7] And here, just as certain accidents of man cause the multiplication of what we call human individuals, so certain accidents of animals multiply the species of animality. Similarly, certain accidents of what is vital cause the multiplication of what is animated and alive. It is no different for certain corporeal accidents which cause the multiplication of corporeality, in the same way certain accidents of the substantial multiply the substance. And finally, in the same way, certain accidents of being cause the multiplication of entity, truth, unity, being, the true, the one.

Fourthly, if you consider the signs and the proofs thanks to which we wish to demonstrate the coincidence of contraries, it will not be difficult to infer that all things are one in the end. All number, be it odd or even, finite or infinite, is reduced to a unity which, repeated in a finite series, posits number, and by an infinite repetition, negates number. You will adopt signs from mathematics and proofs from other moral and speculative sciences. Let us look at signs first: tell me what is more unlike a straight line than the circle? Is there anything more opposite to a straight line than a curve? And yet, they coincide in the principle and the minimum, since

[7] The one substance of Bruno's world is manifest in Poliinnio, and since he belongs to a number of different species, such as man, philosopher, tall, etc., the number and multiplicity which are associated with Poliinnio result from his belonging to various species rather than from the unity of substance which is manifested in him.

(as the Cusan[8], the inventor of geometry's most beautiful secrets, divinely pointed out) what difference could you find between the minimum arc and the minimum chord? Furthermore, in the maximum, what difference could you find between the infinite circle and the straight line? Do you not see that the larger the circle, the more its arc approximates straightness? Who is so blind that he cannot see (fig. 1) how the arc BB, by being larger than the arc AA, and the arc CC, by being larger than the arc BB, and the arc DD, by being larger than the other three, tend to be parts of ever-larger circles, and, therefore, approach ever more closely the straightness of the infinite line of the infinite circle, indicated by IK? We must, therefore, say and believe with absolute certainty that, as that line which is longer is also, because of its greater length, more straight, the longest of all must superlatively be the straightest. The infinite straight line thus finally becomes the infinite circle. Here, then, is how not only the maximum and the minimum converge into one being, as we have already shown elsewhere, but also how, in the maximum and the minimum, contraries come to be but one, and to be indistinct.

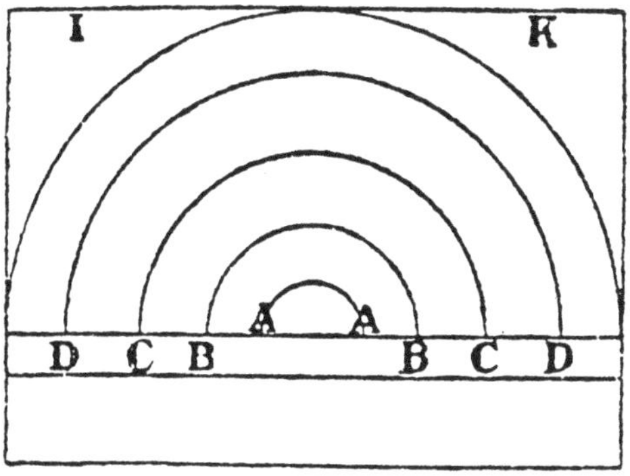

fig. 1.

Moreover, compare, if you will, the finite species to a triangle, since all finite things are seen to participate, by a certain analogy, in the finitude and the limitation of the first finite and first limited thing (just as in all genera, the analogous predicates draw their degree and order from the first and loftiest of the genus), so that the triangle is the first shape which cannot be resolved into another species of simpler shape (while the quadrangle, for instance, can be resolved into triangles), making the triangle the primary foundation of every limited and configurated thing. You will find that the

[8] Nicolas of Cusa, in his *De mathematica perfectione* and *De berillo*.

triangle, as it cannot be resolved into another figure, likewise cannot be composed into triangles whose three angles are greater or smaller, even if the triangles are diverse and varied, of diverse and varied types in terms of greater or lesser size, minimal or maximal. Therefore, if you posit an infinite triangle (I do not mean really and absolutely, since the infinite has no figure; I mean infinite hypothetically, insofar as its angle is useful for our demonstration), it will not have an angle greater than that of the smallest finite triangle, and likewise for that of any intermediate triangle and of another, maximum triangle.

But leaving off the comparison between one shape and another, I mean between triangles, and considering angles, we see that they are all equal, whatever their size, as in this square (fig. 2). This square is divided diagonally into several triangles, and we see that not only are the angles of the

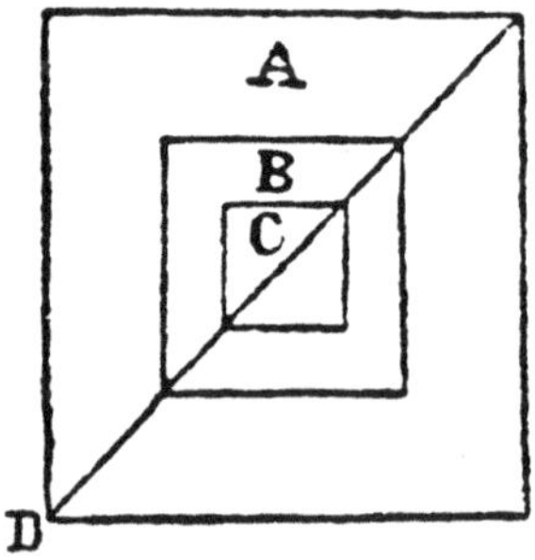

fig. 2.

three squares A, B and C equal, but also that all the acute angles resulting from the division made by the said diagonal, which doubles the series of triangles, are all equal. From this we can very clearly see, by a very marked analogy, how the one infinite substance can be whole in all things, although in some in a finite manner and in others in an infinite manner, in some in lesser measure and in others in greater measure.

Add to this (to see further that, in this one, in this infinity, contraries coincide) that the acute and the obtuse angles are two contraries. But do you not see (fig. 3) that they are formed from a unique, undivided, identical principle, that is, from the inclination made by the line M, which joins perpendicularly the horizontal line BD at point C? Pivoting on point C, and by a simple inclination towards point D, that perpendicular line, that produces, first, two identical right angles and highlights, then, the difference between the acute and the obtuse angle as it approaches point D. When it has reached that point and is united with it, it merges the acute and obtuse angles, which cancel one another out, since the one and the other are united in the potency of one and the same line. The line M, which has been made

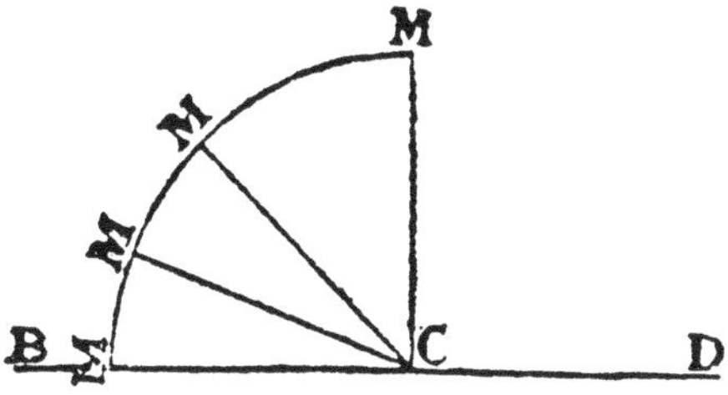

fig. 3.

to unite with and merge with line BD, can, similarly, disunite and separate itself from it, giving rise from the same identical, unique and undivided principle to the most contrary angles, from the maximum acute and the maximum obtuse, to the minimum acute and the minimum obtuse, and thence to their equivalence as right angles, and their merging produced when the perpendicular and the horizontal lines are superimposed.

Let us proceed now to proofs: first, regarding the active primary qualities of corporeal nature, who does not know that the principle of heat is indivisible and, in consequence, is separated from all heat, since the principle cannot be any of the principled things? And if this is true, who can hesitate to affirm that the principle is neither cold nor hot, but that there is one and the same principle for cold and heat? What explains that a contrary is the principle of its opposite, and that, therefore, the transmutations are circular, if not the existence of a subject, of a principle, of a term, and a continuity and a coincidence between the one and its contrary? Are not maximum heat and minimum cold wholly one? Is it not from the limit of maximum heat that we obtain the point of departure in the movement toward cold? It is evident, therefore, that not only do the two maxima sometimes coincide in their opposition and that the two minima coincide in their agreement, but *etiam* [also] that the maximum and the minimum coincide through the vicissitude of transmutation. Therefore, it is not without cause that physicians are often concerned when faced with the best of health, or that those with foresight grow doubly prudent in periods of greatest happiness. Who does not see that corruption and generation derive from the same principle? Is not the end of the corrupt thing the beginning of the thing generated? Do we not similarly say: to take that, is to posit this?

There was that, there is this? If we use our judgement wisely, we see clearly that corruption is nothing but a generation, and generation is nothing but a corruption; love is a hate and hate is a love, in the end. Hate of the contrary is a love of the congruent, and the love of this is the hate of that. Therefore, in substance and at the root, love and hate, amity and discord are the same thing. Where does the physician find the antidote more surely than in poison? Who delivers a better theriac than the viper? The best remedies lie in the worst venoms. Is a potency not the potency of two contrary objects? And how do you think that can be explained, if not because the principle of the being of both the one and the other is one, as is the principle of their conception, and if not because the contraries are related to one and the same substratum, just as they are apprehended by the same sense? Not to mention that the sphere rests on the plane, the concave remains on and settles into the convex, the irascible lives in accord with the patient, the prideful likes the humble the best and the bountiful the miser.

In conclusion, he who wants to know the greatest secrets of nature should observe and examine the minima and maxima of contraries and opposites. There is a profound magic in knowing how to extract the contrary from the contrary, after having discovered their point of union. Poor Aristotle was tending to this in his thought when he posited privation (to which a certain disposition is joined) as the progenitor, parent and mother of form, but he could not get to it. He failed to attain it because, stopping at the genus of opposition, he remained snared by it in such a way that, not having descended to the species of contrariety, he did not reach or even perceive the goal. He strayed completely away from it by claiming that contraries cannot actually concur in the same substratum.

POLIINNIO. You have expatiated in an elevated, rare and exceptional way about the whole, the maximum, about being, principle and the one. But I would like to hear you speak more explicitly about unity, for I find there a *Vae soli* [Woe to the solitary]![9] Moreover, I feel a great anguish at the idea that in my purse and in my wallet, there is but a single coin.

TEOFILO. The unity which is all is not unfolded, nor found in numeric distribution and distinction. It is not a singularity such as you perhaps conceived it, but a unity which is all-embracing and comprehensive.

POLIINNIO. *Exemplum* [An example]? To tell the truth, *intendo,* but *non capio* [I am paying attention, but I do not understand].

[9] Ecclesiastes 4, 10.

TEOFILO. The decade is a unity in the same way, but it is complex; the hundred is no less a unity, but it is more complex, and the thousand is a unity no less than the other two, but much more complex. And what I tell you in arithmetical terms, you must understand in the sense of a greater depth and a greater simplicity as regards the totality of things. The supreme good, the supreme object of desire, the supreme perfection, the supreme beatitude consists in the unity which embraces the whole. We delight in colour; not in a single, express colour, whatever it may be, but above all in the colour which embraces all colours. We delight in sound, not in any particular one, but in a complex sound which results from the harmony of many sounds. We delight in a sensible thing, but we take greatest delight in that which comprehends, in itself, all sensible things; similarly, we take delight in a knowable thing that comprehends all knowable things, an apprehensible thing that embraces all that can be apprehended, a being that embraces everything; we delight, above all, in the one which is itself the all. Just as you, Poliinnio, would prefer the unity of a gem so precious as to be worth all the gold in the world, to the multitude of thousands upon thousands of such pennies as the one you have in your purse.

POLIINNIO. *Optime* [Excellent].

GERVASIO. Here I am grown learned. For if the man who does not understand the one understands nothing, he who really understands the one understands everything. And the closer one gets to the intelligence of the one, the closer one comes to the apprehension of everything.

DICSONO. The same goes for me. If I have understood rightly, I am going away much enriched by the considerations of Teofilo, reliable reporter of the Nolan philosophy.

TEOFILO. Praised be the gods, and may all living things magnify the infinite, perfectly simple, unique, highest and absolute cause, principle and unity.

End of the five dialogues
on Cause, Principle, and Unity

On Magic

On magic

As with any other topic, before we begin our treatise *On Magic*, it is necessary to distinguish the various meanings of the term, for there are as many meanings of 'magic' as there are of 'magician'.

First, the term 'magician' means a wise man; for example, the trismegistes among the Egyptians, the druids among the Gauls, the gymnosophists among the Indians, the cabalists among the Hebrews, the magi among the Persians (who were followers of Zoroaster), the sophists among the Greeks and the wise men among the Latins.

Second, 'magician' refers to someone who does wondrous things merely by manipulating active and passive powers, as occurs in chemistry, medicine and such fields; this is commonly called 'natural magic'.

Third, magic involves circumstances such that the actions of nature or of a higher intelligence occur in such a way as to excite wonderment by their appearances; this type of magic is called 'prestidigitation'.

Fourth, magic refers to what happens as a result of the powers of attraction and repulsion between things, for example, the pushes, motions and attractions due to magnets and such things, when all these actions are due not to active and passive qualities but rather to the spirit or soul existing in things. This is called 'natural magic' in the proper sense.

The fifth meaning includes, in addition to these powers, the use of words, chants, calculations of numbers and times, images, figures, symbols, characters, or letters. This is a form of magic which is intermediate between the natural and the preternatural or the supernatural, and is properly called 'mathematical magic', or even more accurately 'occult philosophy'.

The sixth sense adds to this the exhortation or invocation of the intelligences and external or higher forces by means of prayers, dedications, incensings, sacrifices, resolutions and ceremonies directed to the gods, demons and heroes. Sometimes, this is done for the purpose of contacting a spirit itself to become its vessel and instrument in order to appear wise, although this wisdom can be easily removed, together with the spirit, by means of a drug. This is the magic of the hopeless, who become the vessels of evil demons, which they seek through their notorious art. On the other hand, this is sometimes done to command and control lower demons with the authority of higher demonic spirits, by honouring and entreating the latter while restricting the former with oaths and petitions. This is transnatural or metaphysical magic and is properly called 'theurgy'.

Seventh, magic is the petition or invocation, not of the demons and heroes themselves, but through them, to call upon the souls of dead humans, in order to predict and know absent and future events, by taking their cadavers or parts thereof to some oracle. This type of magic, both in its subject matter and in its purpose, is called 'necromancy'. If the body is not present, but the oracle is beseeched by invoking the spirit residing in its viscera with very active incantations, then this type of magic is properly called 'Pythian', for, if I may say so, this was the usual meaning of 'inspired' at the temple of the Pythian Apollo.

Eighth, sometimes incantations are associated with a person's physical parts in any sense; garments, excrement, remnants, footprints and anything which is believed to have made some contact with the person. In that case, and if they are used to untie, fasten, or weaken, then this constitutes the type of magic called 'wicked', if it leads to evil. If it leads to good, it is to be counted among the medicines belonging to a certain method and type of medical practice. If it leads to final destruction and death, then it is called 'poisonous magic'.

Ninth, all those who are able, for any reason, to predict distant and future events are said to be magicians. These are generally called 'diviners' because of their purpose. The primary groups of such magicians use either the four material principles, fire, air, water and earth, and they are thus called 'pyromancers', 'hydromancers', and 'geomancers',[1] or they use the three objects of knowledge, the natural, mathematical and divine. There are also various other types of prophecy. For augerers, soothsayers and other such people make predictions from an inspection of natural or physical things. Geomancers make predictions in their own way by inspecting mathematical objects like numbers, letters and certain lines and figures,

1 The fourth implied name, 'aeromancers', is not included in Bruno's text.

and also from the appearance, light and location of the planets and similar objects. Still others make predictions by using divine things, like sacred names, coincidental locations, brief calculations and persevering circumstances. In our day, these latter people are not called magicians, since, for us, the word 'magic' sounds bad and has an unworthy connotation. So this is not called magic but 'prophecy'.

Finally, 'magic' and 'magician' have a pejorative connotation which has not been included or examined in the above meanings. In this sense, a magician is any foolish evil-doer who is endowed with the power of helping or harming someone by means of a communication with, or even a pact with, a foul devil. This meaning does not apply to wise men, or indeed to authors, although some of them have adopted the name 'hooded magicians', for example, the authors[2] of the book *De malleo maleficarum* (*The Witches' Hammer*). As a result, the name is used today by all writers of this type, as can be seen in the comments and beliefs of ignorant and foolish priests.

Therefore, when the word 'magic' is used, it should either be taken in one of the senses distinguished above, or, if it is used without qualifications, it should be taken in its strongest and most worthy sense as dictated by the logicians, and especially by Aristotle in Book v of the *Topics*.[3] So as it is used by and among philosophers, 'magician' then means a wise man who has the power to act. Nevertheless, the fact remains that the word, when unqualified, means whatever is signified by common usage. Another common meaning is found among various groups of priests who frequently speculate about that foul demon called the devil. Still other meanings are to be found in the common usages of different peoples and believers.

Given these distinctions, we will deal generally with three types of magic: the divine, the physical and the mathematical. The first two of these types of magic necessarily relate to what is good and best. But the third type includes both good and evil, since the magician may direct it towards either. Although all three types agree on many principles and actions, in the third type, wickedness, idolatry, lawlessness and charges of idolatry are found when error and deception are used to turn things which are intrinsically good into evil. Here, the mathematical type of magic is not defined by the

[2] The authors of this book, first published c. 1486, were Heinrich Kramer (Henricus Institoris) and James Sprenger. An English translation by Rev. Montague Summers has been published under the Latin title *Malleus maleficarum* (New York: Benjamin Blom, 1928; Dover, 1971).

[3] In *Topics*, v, 1–9, Artistotle provides a long list of rules to be used to determine the meaning of words in terms of the properties assigned to things.

usually mentioned fields of mathematics, i.e., geometry, arithmetic, astronomy, optics, music, etc., but rather by its likeness and relationship to these disciplines. It is similar to geometry in that it uses figures and symbols, to music in its chants, to arithmetic in its numbers and manipulations, to astronomy in its concerns for times and motions, and to optics in making observations. In general, it is similar to mathematics as a whole, either because it mediates between divine and natural actions, or because it shares or lacks something of both. For some things are intermediates because they participate in both extremes, others because they are excluded from both extremes, in which case they should not be called intermediates but a third species which is not between the other two but outside of them. From what has been said, it is clear how divine and physical magic differ from the third type.

To turn now to the particulars, magicians take it as axiomatic that, in all the panorama before our eyes, God acts on the gods; the gods act on the celestial or astral bodies, which are divine bodies; these act on the spirits who reside in and control the stars, one of which is the earth; the spirits act on the elements, the elements on the compounds, the compounds on the senses; the senses on the soul, and the soul on the whole animal. This is the descending scale.

By contrast, the ascending scale is from the animal through the soul to the senses, through the senses to compounds, through compounds to the elements, through these to spirits, through the spirits in the elements to those in the stars, through these to the incorporeal gods who have an ethereal substance or body, through them to the soul of the world or the spirit of the universe; and through that to the contemplation of the one, most simple, best, greatest, incorporeal, absolute and self-sufficient being.

Thus, there is a descent from God through the world to animals, and an ascent from animals through the world to God. He is the highest point of the scale, pure act and active power, the purest light. At the bottom of the scale is matter, darkness and pure passive potency, which can become all things from the bottom, just as He can make all things from the top. Between the highest and lowest levels, there are intermediaries, the higher of which have a greater share of light and action and active power, while the lower levels have a greater share of darkness, potency and passive power.

As a result, all light in lower things, which comes to them from above, is more powerful in higher things. And also, all darkness in higher things is stronger in lower things. But the nature and power of light and darkness are not equal. For light diffuses and penetrates through the lowest and deepest darkness, but darkness does not touch the purest sphere of light. Thus, light penetrates and conquers darkness and overflows to infinity, while darkness does not penetrate or overwhelm or equal the light, but rather is very weak compared to light.

Parallel to the three types of magic mentioned above, there are three different worlds to be distinguished: the archetypal, the physical and the rational. Friendship and strife are located in the archetypal world, fire and water in the physical world, and light and darkness in the mathematical world. Light and darkness descend from fire and water, which in turn descend from peace and strife. Thus, the first world produces the third world through the second, and the third world is reflected in the first through the second.

Leaving aside those principles of magic which play on the superstitious and which, whatever they be, are unworthy of the general public, we will direct our thoughts only to those things which contribute to wisdom and which can satisfy better minds. Nevertheless, no type of magic is unworthy of notice and examination, because every science deals with the good, as Aristotle says in the introduction to his *De anima*,[4] and as Thomas and other more contemplative theologians agree. Nevertheless, all this should be kept far away from profane and wicked people and from the multitudes. For nothing is so good that impious and sacrilegious and wicked people cannot contort its proper benefit into evil.

In general, there are two types of efficient cause: nature and the will. The will is threefold: human, spiritual and divine. Nature, as used here, is twofold: intrinsic and extrinsic. Furthermore, intrinsic nature is of two kinds: matter or the subject, and form with its natural power. Extrinsic nature is also of two kinds: the first, which is preferably called an image of nature, is a trace and shadow or light which remains in a thing within its body, like light and heat in the sun and in other hot bodies; the second emanates and radiates from a subject, like light, which flows from the sun and is found in illuminated things, and like heat, which resides with light in the sun and is also found in heated bodies.

By examining the number of these causes, we can descend to the differentiation of powers or of effects produced by the first cause through the intermediate causes down to the closest and lowest ones, by restricting the universal cause, which, itself, does not attend to any one available

[4] Aristotle, *De anima*, I, I (402.a.1–2).

subject rather than another. For, although this cause and its causal power always remain immutably the same, it produces contrary (and not just different) effects in different subjects with the help of different types of matter. As a result, there need be only one such simple, principal efficient cause, like there is only one sun, one heat and one light, which by turning forward or backward, by approaching and receding, mediately and immediately causes the winter, the summer, and their different and contrary weather, and the ordering of the seasons.

Matter is also derived from this same cause, if we wish to believe those who think that the four commonly mentioned elements change into each other. The originator of this view was Plato, who sometimes says that all things were produced from one matter and one efficient cause. But whatever may have been the method of production used by the first universal cause, and whether one assumes only one or many material principles, any human or spiritual secondary cause must recognize that, because of the great multitude and variety of producing species, there are many types of matter having act or form, through which a subject is able to influence things outside of itself.

In regard to the powers or forms or accidents which are transmitted from subject to subject, some are observable, for example, those that belong to the genus of active and passive qualities, and the things that immediately follow from them, like heating and cooling, wetting and drying, softening and hardening, attracting and repelling. Others are more hidden because their effects are also obscure, for example, to be happy or sad, to experience desire or aversion, and fear or boldness. These are said to be caused by external impressions acting on the cognitive power in humans and on the estimative power in animals. Thus, when a child or infant sees a snake, or when a sheep sees a wolf, it conceives an image, without any other experience, of hostility and of fear of its own death or destruction.

The explanation of this is to be found in the internal sense, which is, indeed, moved by the external impressions, although indirectly. For nature gives not only existence to each species, but also the desire in each individual to preserve itself in its present state. Thus, it implants in each thing an internal spirit, or sense, if you prefer that word, by which as from an internal dictate it recognizes and avoids great dangers. This can be seen not only in the examples given above, but also in all things in which, even if they seem to be defective or dead, there still resides a spirit striving with all its power to conserve the present condition. This happens in falling drops of liquids which form a sphere to avoid disintegration, and also in falling bodies which are attracted to a centre and tend to gather their parts into a sphere lest they break up and disperse. This is also found in pieces of straw or wood thrown on a fire, and in thin tissues and membranes, which recoil to avoid, somehow, their own destruction.

This particular sense is located in all things and is a form of life, although, in accordance with the common custom, we do not call it an animal, which refers to a specific soul, because these components cannot be called animals. Nevertheless, in the order of the universe, one can recognize that there is one spirit which is diffused everywhere and in all things, and that everywhere and in all things there is a sense of grasping things which perceives such effects and passions.

Just as our soul produces, originally and in a general way, all vital activities from the whole body, and even though the whole soul is in the whole body and in each of its parts, nevertheless, it does not produce every action in the whole body or in each of its parts. Rather, it causes vision in the eyes, hearing in the ears and taste in the mouth (but if the eye were located in any other place, we would see in that place, and if the organs of all the senses were located in any one place, we would perceive everything in that place). In the same way, the soul of the world is in the whole world, and is everywhere so adapted to matter that, at each place, it produces the proper subject and causes the proper actions. Therefore, although the world soul is located equally everywhere, it does not act equally everywhere, because matter is not arranged to be equally disposed to it everywhere. Thus, the whole soul is in the whole body, in the bones and in the veins and in the heart; it is no more present in one part than in another, and it is no less present in one part than in the whole, nor in the whole less than in one part. Rather, it causes a nerve to be a nerve in one place, a vein to be a vein elsewhere, blood to be blood, and the heart to be the heart elsewhere. And as these parts happen to be changed, either by an extrinsic efficient cause or by an intrinsic passive principle, then the activity of the soul must also change.

This is the most important and most fundamental of all the principles which provide an explanation of the marvels found in nature; namely, that because of the active principle and spirit or universal soul, nothing is so incomplete, defective or imperfect, or, according to common opinion, so completely insignificant that it could not become the source of great events. Indeed, on the contrary, a very large disintegration into such components must occur for an almost completely new world to be generated from them.

While bronze is more similar to gold and is closer to the distinctive properties of gold than are the ashes of bronze, still, in a transmutation, these ashes of bronze are closer to the form of gold than is bronze. Likewise, we see that all seeds, which are oriented to producing a particular species, happen to be rather alike as though they were of the same, and not different, species, since they are similar and distinctive and related. He who believes otherwise is like someone who thinks that an ape can be changed into a human more easily than the seed, implanted in a woman, which previously was food and bread.

Nevertheless, in every production, there must be present a similarity and a form of the same species. Just as a house or a garment results from a model in the maker's mind in the case of artefacts, likewise, in the productions of nature, a species of things is generated and defined by the exemplar, which is distinctive of the matter which generates the form. For example, we see the same types of food, and the same heavens, water and houses reproduced in substance: a dog into a dog, a human into a human, a cat into a cat. And a dog generates the same species of dog, and a human the same species of human.

From this, it is clear that the entire cause of the differences is due to an idea, which is generally present everywhere in nature, and which is later limited to this or that species, depending on whether one or other species resembles the idea more. As a result, any magician who wishes to carry out his work in accordance with nature must especially understand this ideal principle and how it applies specifically to species, numerically to numbers and individually to individuals. From this, he formulates an image and the proportions of the matter so formed, and with good reasons reinforces the result with the wisdom and power of his magic. Many also bring about cures and injuries by connecting symbols with particular components or by appealing to those who communicate with or take part in curing or destructive forces. In this way, the work of magic is restricted and applied to a particular individual.

Leaving aside other arguments, it is clear from these experiences that every soul and spirit has some degree of continuity with the universal spirit, which is recognized to be located not only where the individual soul lives and perceives, but also to be spread out everywhere in its essence and substance, as many Platonists and Pythagoreans have taught. As a result, vision grasps the most distant things immediately and without motion, and, indeed, the eye, or some part thereof, extends immediately to the stars or immediately from the stars to the eye.

Furthermore, the soul, in its power, is present in some way in the entire universe, because it apprehends substances which are not included in the body in which it lives, although they are related to it. Thus, if certain impediments are excluded, the soul has an immediate and sudden presence with the most distant things, which are not joined to it by any motion, which nobody would deny, but rather are directly present in a certain sense.

Experience teaches this also in the case of those whose nose has been cut off; if they arrange to grow a new nose for themselves from the flesh of some other animal, and if that animal whose flesh was used dies, then as the body of that animal rots, so does the borrowed nose. From this, it is clear that the soul diffuses outside of the body in every aspect of its nature.[5] It also follows that the soul knows not only the members of its own body, but also everything for which it has any use, participation, or interaction.

There is no value in the stupid argument, advanced by those who lack true philosophical principles, that a thing which is touched by something else does not itself perceive that. Indeed, in one sense, this is true if we are distinguishing between species or individuals, but it is false if we are distinguishing one bodily part from another. For example, if someone injures a finger or pricks one part of the body with a pin, the whole body is immediately disturbed everywhere and not just in that part where the injury occurred.

And so, since every soul is in contact with the universal soul, it is not possible to find in this case the same effects which occur in bodies which do not mutually penetrate into each other. Rather, for spiritual substances, a different comparison is needed. For instance, if innumerable lamps are lit, they all act together as though they were one light, and no one light impedes or reflects or excludes another. The same thing happens when many voices are diffused through the same air, or if many visible rays, to use the common saying, spread out to reveal the same visible whole. All these rays pass through the same medium, and while some move in straight lines and others obliquely, they do not interfere with each other. In the same way, innumerable spirits and souls, when spread out through the same space, do not interfere with each other such that the diffusion of one would affect the diffusion of an infinity of others.

This power belongs not only to the soul but also to certain accidents, like sound, light and vision. The reason is that the whole soul is located in the whole body and in every part of the body, and that the whole soul

[5] This is Thesis XV in Bruno's *Theses de magia*. See his *Opera latine conscripta*, III, 466.

apprehends all things, however diverse and distant, which are around it outside of its body. This is a sign that the soul is not included in the body as its first act and substance,[6] and that it is not circumscribed by the body. Rather, in itself and by itself, it should be understood only as a second act. This principle is the cause of innumerable marvellous effects, although its nature and power need to be investigated. This soul and divine substance cannot be inferior to the accidents which issue from it as its effects, traces and shadows. I declare that if the voice operates outside the body which produces it, and enters as a whole into innumerable ears on all sides, then why cannot the whole substance, which produces the voice which is tied to certain organs of the body, be located in different places and parts?

Furthermore, it must be noted that occult intelligence is not heard or understood in all languages. For the voices spoken by humans are not heard in the same way as the voices of nature. As a result, poetry, especially of the tragic type (as Plotinus says), has a very great effect on the wavering thoughts of the soul.

Likewise, not all writings have the same impact as those markings which signify things by the particular way in which they are drawn and configured. Thus, when certain symbols are arranged in different ways, they represent different things: in a circle, the attraction of love; when opposed, the descent into hate and separation; when brief, defective and broken, they point to destruction; when knotted, to bondage; when strung out, to dissolution. Furthermore, these symbols do not have a fixed and definite form. Rather, each person, by the dictate of his own inspiration or by the impulse of his own spirit, determines his own reactions of desiring or rejecting something. And thus, he characterizes for himself each symbol according to his own impulse, and as the divine spirit personally exerts certain powers which are not expressed in any explicit language, speech, or writing.

Such were the figures, so well designed by the Egyptians, which are called hieroglyphics or sacred symbols. These were specific images selected from natural objects and their parts to designate individual things. The Egyptians used these symbols and sounds to converse with the gods to accomplish extraordinary results. Later, when Theuth,[7] or someone else, invented the letters of the type we use today for other purposes, this

[6] This sentence is an explicit rejection of Aristotle's definition of the soul in *De anima*, II, 1 (412.a.28–30).

[7] See Plato, *Phaedrus*, 274 c–e, for an account of this legend about the origins of written language.

resulted in a tremendous loss, first of memory, and then of divine science and magic.

Like those Egyptians, magicians today formulate images, written symbols and ceremonies, which consist of certain actions and cults, and through which they express and make known their wishes with certain signals. This is the language of the gods which, unlike all other languages which change a thousand times every day, remains always the same, just as natural species remain always the same.

For the same reasons, the spirits speak to us through visions and dreams, but we claim that these are enigmas, because of our unfamiliarity and ignorance and weak capacities, even though they are the very same sounds and very same expressions used for representable things. Just as these sounds elude our grasp, likewise our Latin, Greek and Italian sounds sometimes fail to be heard and understood by the higher and eternal spirits, which differ from us in species. Thus, it is no easier for us to be able to communicate with the spirits than it is for an eagle to converse with a human. Just as there can be conversation and agreement only by means of gestures between two groups of humans who do not share a common language, likewise, there can be communication between us and certain types of spirits only by the use of certain signs, signals, figures, symbols, gestures and other ceremonies. The magician, especially when using the kind of magic which is called 'theurgy', can hardly accomplish anything without such sounds and symbols.

On the communion and interaction of things

From the above, one can understand and explain how interaction occurs not only between things which, to the senses, are near each other, but also between things which are far apart. For, as was said above, things are united by a universal spirit which is present as a whole in the whole world and in each of its parts. As a result, just as different lights come together in the same space, likewise, the souls of the universe, whether they be finite or infinite in number, interact in their powers and activities. However, this does not happen to bodies, because they are limited and circumscribed by their surfaces and surroundings, and because they are composed of innumerably different parts in different bodies and places (if we can speak of place rather than space.) Therefore, no body can act on another body, and no matter on other matter, nor can the material parts of one body act on the parts of another body, but rather, all action comes from quality and form and ultimately from soul. The soul first changes the dispositions, and then the dispositions change bodies. Thus, bodies act on distant bodies, on nearby ones and on their own parts, by means of a certain harmony, joining and union which comes from form. For, since every body is governed by a soul or a spirit which connects its parts, and since one soul acts on another nearby soul in any direction and wherever it is, it follows necessarily that a soul moves that body, wherever it is, because it is controlled by, and subordinate to, that soul. Whoever is aware of this indissoluble continuity of the soul and its necessary connection to a body will possess an important principle both to control natural things and to understand them better.

From this follows clearly the reason why a void, i.e., a space empty of any body, does not exist. For no body can leave one place without being replaced by another. During life, the soul does leave its own body, but it cannot leave the universal body, nor can it be abandoned by the universal body, if you prefer to state it that way. For, when it leaves one simple or complex body, it moves into another simple or complex body, or from one body left behind, it goes to and enters into another. Thus, it has an indissoluble connection with universal matter. And since its own nature is to be a continuous whole everywhere, we realize that it exists together with a material body everywhere. From this, we conclude that the void does not exist in the sense of a space with no matter in it, but rather a void is a space in which different bodies move and succeed each other. It also follows that the motion of the parts of one body towards the parts of another body is continuous. Motion occurs through a continuous space which is not interrupted by any void located between full spaces, unless we wish to say that a space in which there is no sensible body is a void.

A continuous body is an unobservable body, that is, an airy or ethereal spirit. It is very active and very powerful and very similar to the soul, and it is quite different from the dense sluggishness found in observable, composite substances. The above mentioned powers of unobservable spiritual bodies are the source of all the powers in observable bodies themselves. Indications of this are the airy spirit which agitates and embroils all the seas, and the invincible force of the winds which, even when they are rather calm and quiet, disturb the earth, break trees and destroy houses. Lucretius put it well when he said that it is this spiritual body which performs all actions in observable things.[8] Many philosophers have thought that this

[8] Lucretius, *De rerum natura*, v, 235ff.

spiritual body is the same thing as the soul, and the poet has said that this air 'has the power of fire and the soul'.[9]

Furthermore, fire, which does not consist of dense matter like coal, which is rather an ignited body, is to be understood as different from air only accidentally. For true fire is, indeed, a spirit, which is sluggish when it is located in an ignited body, but it is lively when it is outside of an ignited body, and it is a form of motion in a flame or in some intermediate state.

This spirit, acting in different ways, forms different bodies and animals. And although not all composite bodies are animals, it must be noted that all animals have a soul which is of the same type in all the members of one genus. However, that soul is not actuated in one and the same way because of differences in the dispositions of matter and in objective ideals. From this, it follows that, since there are different and contrary forms, there are also differences and reasons why some animals congregate with each other, being attracted and impelled to various places, while others flee from or pursue each other. All of this is due to the way they are structured.

All things desire to preserve their own existence, and thus they forcibly and unwillingly resist separation from the place where they exist and persevere. This force is so strong that the sun or fire attracts water to itself through airy space only after it has first made the water like air, that is, by converting it into a vapour. After that has happened, then the substance which was water is attracted willingly, and by means of the same impulse of attraction, it tries to adapt itself so that it slowly becomes more and more like fire, and finally, it becomes fire itself. On the other hand, that most subtle body which is contained in the spirit in the form of fire changes back into water by the opposite sequence as it congeals and thickens.

Therefore, the same substance and matter changes from water to vapour, from vapour to air, and from air to the thinnest and most penetrating ethereal body. The latter has been called a 'spirit' by the Egyptians, Moses and Dionysius of Apollonia, although they differ because Moses did not distinguish spirit from soul (according to his words; I do not judge his meaning), which the others did. The other substance [earth] is dry and composed of atoms, which are very solid and indissoluble bodies. In themselves, they are neither continuous nor divisible and, thus, cannot be changed into any other body. And the substance of water or spirit or air, which is the same, never changes into the substance of atoms or dry earth, nor *vice versa*.

[9] Virgil, *Aeneid*, VIII, 403.

This philosophy is supreme, divine and true, since it is quite in agreement with nature by positing the following principles of reality: first, water or the abyss or the Styx; second, dryness or atoms or earth (I am not speaking of the terrestrial globe); third, spirit or air or soul; fourth, light. These are so different from each other that one cannot be transformed into the nature of another, although they do come together and associate, sometimes more or less, sometimes all or some of them.

On the double motion of things and on attraction

There are two kinds of motion, natural and preternatural. Natural motion comes from an intrinsic principle, while preternatural motion is from an extrinsic principle; natural motion is in harmony with the nature, structure and generation of things, preternatural motion is not. The latter is twofold: violent, which is against nature, and ordered or structured, which is not contrary to nature. What is commonly called natural motion is found in all genera or in all the categories, leaving out the distinction between motion and mutation.[10]

For now, we omit all the other classes of motion and their species and consider only natural motion in place. One type of this is possessed by natural things and does not move a thing away from it own proper place; this is circular motion, or a version thereof. The other type is straight line motion, which is not naturally possessed by natural things. For example, air moves in a straight line to fill a void. A stone moves through air, and a body which is heavier than water moves through water, in a straight line in order to occupy the place in which it either is at rest or moves naturally. And as much as it can, a contrary flees from its contrary in a straight line, for example, fumes, vapour and water from fire (for it goes faster to a greater distance through a straight line). Likewise, similar and agreeable things tend towards each other in a straight line, for example, straw to amber, and iron to a magnet, so that they can rest together or move better and more easily.

There is also a third type of local motion, which is an inflow and an outflow found in all natural things when any of their parts are ejected in various ways and in every direction. For now, we will call this 'spherical' motion. For it does not occur either in a straight line, or to or from

[10] This is an Aristotelian distinction in which 'motion' refers to changes in quality, quantity and place, while 'mutation' refers to changes in substance or essence.

or around a centre. Rather, it occurs along an infinity of lines from the same centre, for while some parts of the body are ejected and emitted outward from the body's convex surface and perimeter, others are reciprocally received and absorbed. Bodies grow and are invigorated when the inflow of beneficial things is greater than their outflow, and they age and weaken and become sluggish when the inflow of extraneous things is greater than the outflow of natural things. This is the reason why corruption and change occur in things, including all changes or alterations and disintegrations ...[11]

There is no controversy over the evidence for the first two types of motion, and, as a result, the understanding and classifications of them are well known. But a more careful consideration of the third type will be found to be not only needed and helpful, but also necessary. The situation is especially clear in things which have very strong sensible qualities. For instance, fire warms in every direction and not just on one side or another. For, as soon as it is lighted, it sends out its light and flames in every direction. Likewise, a sound and a voice penetrate equally in all directions, if they occur in a medium which is open on all sides. In the same way, it is quite clear that the sense of smell is activated by the continuous emission of small parts from an odoriferous body. This could not happen unless that body's substance were to flow out and emit its parts in all directions. The same thing happens in the case of reflections and other such observable occurrences. Innumerable other accidental things are caused by certain parts flowing out, and sometimes these parts travel an enormous distance from a very small observable source, as is clear when a small amount of something emits a smell for many years.

In addition to these observable qualities or powers which are emitted from bodies spherically, there are other, more spiritual and less heavy ones which act not only on the body and on the senses, but also on the interior spirit. The more powerful ones touch the powers of the soul and cause various effects and passions, as is commonly thought to be true of many stones, herbs and minerals. This is also clear in fantasies and in cases where the eye has actively or passively been hit. An example is the basilisk[12] who, by looking at a man a long way off, can kill him with the sharpness of its vision.

[11] There is a break in Bruno's text at this point.

[12] The basilisk, also called a cockatrice, was a legendary monster resembling a lizard which was said to have been hatched by a snake from a cock's egg. Its breath and glance were supposedly fatal.

How a magnet attracts iron, coral attracts blood, etc.

What was said above explains why magnets naturally attract things. There are two kinds of attraction. The first occurs by agreement, as when parts move to their proper place and are oriented to that place, and when similar and harmonious things attract each other. The second type occurs without agreement, as when contraries come together because the one which cannot escape is overwhelmed, as when moisture is attracted by fire. This is clear in the case of a burning object being held above a bowl containing water, where the water is sucked up by the heat and rises rapidly. The same thing happens when waterspouts and hurricanes occur at sea, with the result that sometimes even ships are thrown a great distance upwards by the waves.

Attraction occurs in three ways. The cause of the first type is clear to the senses, as is shown in the cases mentioned above. This also happens when the attraction and absorption of air attract objects contained in the air. This is, likewise, evident in pipes through which water is sucked, and thus rises to any level. This happens for the reason given. For if the air in the tube is attracted, and if there is no other air to take its place, then water or earth or something else will fill that space. If nothing can replace it, then the air would be held back and retained by the power of the vacuum, as is clear when an opening is obstructed by objects being sucked in and swallowed. Another example occurs when the tongue and lips are held together and their opening is very tightly compressed around the mediating air, and *vice versa*, when one sucks so that there is elicited from the mouth's pores a spirit which restores and re-establishes what had been removed from its proper place or space.

There is another type of attraction which is not perceived by the senses. This is the case of a magnet attracting iron. The cause of this cannot be attributed to a vacuum or to any such thing, but only to the outflow of atoms or parts, which occurs in all bodies. For when atoms of one type move towards and mutually encounter other atoms of a similar type or of a congenial and compatible nature, the bodies develop such an attraction and impulse for each other that the overpowered body moves towards the whole of the stronger body. For since all the parts experience this attraction, then so must the whole body also be attracted.

This is illustrated for the senses in the case of two burning lamps. When the lower one is extinguished, fumes and spirits flow up from it (these are well disposed to become flames or to be nourishment for fire), and the upper flame then rapidly descends to re-ignite the lower lamp. This is also found in the small flames of torches. To avoid being extinguished by moisture absorbed from the environment, they clearly are attracted to flammable materials located nearby, and they clearly are attracted to a larger flame either in a straight line or indirectly by jumping or leaping across.

And thus it happens that the overpowered parts of iron are attracted by some type of power or quality (although not all activities in such natural, composite things are due to active and passive qualities); even though this happens sometimes by necessity, that is not the rule. The fact that this attraction is caused by the outflow of parts from such bodies also indicates that when a magnet or amber is rubbed, it attracts iron or straw much more strongly. For the heat causes more parts to be emitted, since it opens the pores and rarefies the body.

From this, it is clear that a similar explanation is to be given of how rhubarb attracts choleric humours from the extremities and surface parts of animals to their intestines, when it has sufficient power, that is, when it is not so strong as to be expelled by nature before it acts, nor so weak as only to move the humours and not attract them.

In magnets and similar things, the attractive force and power is not due to an active or passive quality, in the commonly-used sense of a kind of action or passion, as is found in the four elements. A sign of this is that when a piece of iron touches a magnet, it acquires the same power of attracting other pieces of iron. This could not happen if this were due to an elementary quality. For when heat and coldness are accidentally present in a subject, they quickly disappear when the source of heat is removed. Therefore, one must explain this in terms of the emission of parts or of a spiritual substance which flows from the magnet into the iron. It is difficult to imagine any other or even a similar cause of these effects. Also from this perspective, which is fully self-consistent, it is easy to evaluate the various fantasies and dreams which others have mentioned as the causes of this attraction.

This same explanation and cause accounts for the fact that diamonds are said to block such an attraction, and similar types of explanation account for various other things. For the outflow of a specific power can weaken another power, or actuate and sharpen certain other powers. Thus, it is said that diamonds confer magnanimity on those who wear them.

It is not easy to explain why magnetic attraction occurs at the pole of the earth, especially if what some say is not true, namely, that in that region there are many large magnetic mountains. This is very hard to believe, but let us assume it anyhow. Then we ask why this attraction occurs at all distances from that place. We are not speaking here of active but of passive magnetic attraction. I have not yet experienced whether a magnet attracts another magnet. If there are such mountains, and if they exert their power at such great distances, then since they attract a compass needle at the equator and in the tropics, in our region of the earth they would attract men wearing armour. But this is completely ridiculous. If we grant that a magnet attracts iron when there is nothing except air between them, and if the attraction occurs in a straight line from our region of the earth to those places across the northern sea where the magnetic mountains and cliffs are located, then a large arc of the earth lies in between. Therefore, the magnet would attract iron (or another magnet, if that could happen for a similar reason), if our magnet is located at A and the mountains are at D. The

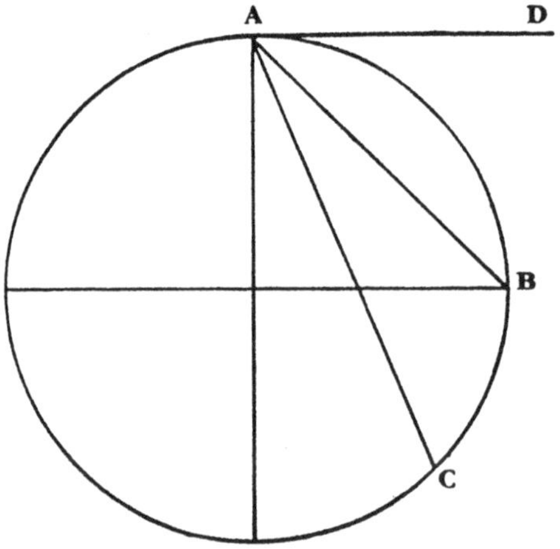

attraction must then occur at B or C, and thus, it crosses either through the large straight line distance AB or AC as indicated, or through the large arc of the earth AB and AC as indicated.

It is clear that this common and well-known argument falls apart for many reasons. To these objections we add that these magnetic mountains do not have the power to attract a magnet because they are similar to each other; if that were the case, we would see a very small magnet attracted by a very large one. Thus, we cannot appeal to attraction as the cause of this effect, because, as was said, it is iron which is attracted, while the magnet rather moves away, for this mineral and iron, which are derived from the earth and are cold like the earth, are contraries by nature. What happens to them is the opposite of what happens to almost all flowers which turn towards the sun and follow the path of the sun, as can be seen, for the sake of this argument, not only in heliotropes, but also in the narcissus, the crocus and in innumerable other flowers. Therefore, we can safely say that these things, which are hostile to the sun and to heat, turn towards and hurry towards those places which are the most removed from the sun and heat.

Epilogue on the motions which occur in things

Thus, we find that local motion occurs in many ways. First is the motion which constitutes and preserves life (i.e., the circular motion of things in their own place due to the soul, or native spirit, as was said above); second is the fleeing of a contrary; third is the acquisition of something helpful or good; fourth is the expulsion and rejection by a contrary impulse; fifth is the violent attraction by a contrary which needs or seeks some material to convert into itself; sixth is an animal choice in accordance with the tendency of a natural power; the last is a violent motion, which either impedes or stops natural powers by some device or desire, or which is due to nature itself which, while it is strong enough to move something in one direction, blocks or impedes another motion of a lesser power, as happens almost everywhere. For example, a natural flow of water in one direction is stopped by another flow of water, as occurs when rivers flowing into the ocean are resisted by the flow of the sea and are turned back for many miles towards their sources.

On the bonding of spirits

As was said above, some spirits reside in more subtle matter, others in more dense matter; some reside in composite bodies, others in more simple bodies; some in observable bodies, others in unobservable bodies. As a result, the operations of the soul are sometimes easier, sometimes more difficult, sometimes weaker, sometimes well adapted, sometimes impossible. Some spirits operate within one genus, others act more efficaciously in another genus. Thus, humans possess certain operations and actions and desires not found in demons, and *vice versa*.

It is easy for demons to penetrate through bodies and to initiate thoughts in us. The reason for the latter is that they convey certain impressions directly to our internal senses, just as we ourselves sometimes seem to think of something suggested by the internal senses. This knowledge seems to occur according to the following comparison and analogy. If one wishes to generate a thought in someone standing at a distance, one must shout so that the thought is produced in their internal sense through their hearing it. But if the person is closer, a shout is not needed, only a quieter voice. And if the person is immediately nearby, a whisper in the ear suffices. But demons have no need of ears or voices or whispers because they penetrate into the internal sense directly, as was said. Thus, they send not only dreams and voices and visions to be heard and seen, but also certain thoughts which are hardly noticed by some. They communicate truths sometimes through enigmas, and sometimes through sense impressions. Sometimes they may even deceive. Not all things are granted to everyone, although they always happen in a definite sequence and order.

Not all spirits or demons have the same level of existence, power and knowledge. Indeed, we know that there are many more species of them than there are of sensible things. Thus, some of them are brute animals and cause injury without any reason. Although these are far inferior to humans in knowledge, they still can do as much harm as dangerous animals or poisons. Mark called these spirits 'deaf and dumb',[13] i.e., they are without reason, since they recognize no commands, and they do not hear or perceive any threats or prayers. As a result, it has been declared that it is impossible to banish them, although it has also been said that they can be controlled and conquered by fasting or abstinence, and by prayer or lofty meditation, and by the power of the senses. This latter is a medical issue, for spirits of this crass type, like food and pleasure, are located in our dense humours and earthy melancholies, which a doctor controls either by thinning them out through fasting or by expelling them with a proper dose of laxative.

There is another type of demon which is fearful, suspicious and credulous. These hear and understand voices but do not distinguish the possible from the impossible, or the appropriate from the inappropriate. They are like humans who are dreaming and disturbed by fantasies. This type of demon is usually expelled by threats of death, prison, fire and other such things.

There are other, wiser demons which reside in pure air, which is a simple substance. They are affected by no cult, no religious practice and no

[13] Mark, 9:25.

prayers. Rather, they freely distort all these things and play with humans by counterfeiting illusions of fear, anger, religion and such things. They understand languages and the sciences, but never make any firm assertions. And so these hateful demons introduce confusion and doubt into the human mind and senses.

There are also ethereal spirits which are pure and luminous. All agree that they are hostile to no one and are completely good and friendly to virtuous men. But the airy spirits are friendly to some, and hostile and hateful to others.

The aqueous and terrestrial spirits are hostile, or at least are not friendly, since they are less rational and more fearful. In accordance with the saying, 'They hate what they fear,'[14] they deliberately cause injury.

But the spirits of fire, which are more properly called heroes and gods, are said to be the ministers of God. The cabalists call them Fissim, Seraphim and Cherubim, and the prophet of the Psalms said, 'He made the winds to be his angels, and the flames of fire his ministers.'[15] Hence, Basil and Origen rightly argue that the angels are not completely incorporeal, but are spiritual substances; that is, they are animals with very subtle bodies, which divine revelation has said are fire and flames of fire.

In every group of spirits there are sovereigns and rulers, ministers, leaders, governors and ranks, by which the wiser and more powerful dominate and direct the more ignorant and more uncultured. These roles do not endure forever, but they are also not as briefly constituted as they are among humans. For in many ways, the lives of spirits are not comparable to ours, since the soul's union with a simple body is much more easily maintained than it is with bodies like ours, which are composed of contraries. Their bodies very easily ward off change. Thus, air and water undergo less change than do composite bodies. Furthermore, they are easily restored. For example, when air is divided, it is reunited very easily, and portions of water reunite after they have been separated. Thus, Virgil did not use a ridiculous poetic figure of speech when he said that Aeneas frightened the shades when he cut through their abode with his drawn sword.[16]

These various spirits occupy the bodies of humans, animals, stones and minerals. There is no body which is completely devoid of spirit and intelligence. Furthermore, no spirit possesses a permanent location for itself. Rather, spirits fluctuate from one matter to another, and matter fluctuates from one spirit to another, and from one nature or composition to another.

[14] Cicero, *De officiis*, II, 7, 3. [15] Psalm, 103:4. [16] Virgil, *Aeneid*, VI, 260, 290, 388.

This is what alteration, mutation, passion and even corruption are: namely, the separation of certain parts from others, and their recombination with still others. For death is nothing more than such a disintegration. No spirit and no body ever perishes; rather, there is only a continual change of combinations and actualizations.

Parallel to the various actualizations, which arise from the various compositions of things, there are various loves and hates. As was said, everything desires to remain in its present state of existence and does not comprehend, or even think about, any other new state of being. Therefore, there is, in general, a bonding of reciprocal love of a soul for its own body, and of that body (in its own way) for its soul. Thus, the diversity of natures and drives gives rise to a variety of bonds which affect both spirits and bodies. We will discuss these bonds immediately after we have first defined the analogy between spirits and composites.

On the analogy of spirits

Porphyry, Plotinus and the other Platonists assign bodies to spirits as follows. The best and purest spirits, which are also called 'gods', have bodies of fire, which is the purest and simplest substance. The spirits which have denser elementary bodies exist only by sharing in a more subtle element. Thus, airy spirits have bodies mixed with air and fire; aqueous spirits have bodies mixed with air and fire; terrestrial spirits have bodies mixed with water, air and fire. These substances are invisible because of their thinness. Furthermore, terrestrial and aqueous spirits sometimes choose to make themselves visible by means of dense and concrete vapours, and they appear in the purer regions, where the air is more calm and quiet.

I, myself, have seen them at Mount Libero and at Mount Lauro. And they have appeared not just to me, but frequently to the local inhabitants to whom they are sometimes hostile (but only moderately so), by stealing and hiding the local animals, which they later return in a few days to their stables.

It is well-known and widely accepted as true that these spirits have also frequently appeared to workers in gold mines and in other underground places, for example, in the mountains of Gebenna. These spirits sometimes harmed them, sometimes helped them and sometimes predicted events. This same type of spirit is found near Nola in a desolate place near the temple of Portus, and under a certain cliff at the foot of Mount

Cicada,[17] which was once used as a cemetery for plague victims. I myself, as well as many others, have personally experienced them while walking through that place at night. I was bombarded with many stones which violently exploded at a very short distance from my head and other parts of my body, and this continued in a threatening manner for some time. Nevertheless, these stones did not inflict any bodily injuries on me or on any of the others who reported the same experience. These incidents are reported by Psellus in his book *De daemonibus* (*On Demons*), where he describes them as refugees from light and as throwers of stones, although their projectiles are harmless.

The existence of subterranean demons is established not only by the senses, experience and reason, but also by divine authority in the very wise Book of Job, which contains a great deal of the most profound philosophy. When Job curses the day he was born with the words, 'May the day of my birth perish', he adds after a few sentences, 'Why was light given to one in misery, and why was life given to those who are bitter of heart?' 'Why did I not perish as soon as I left the womb?' 'Why was I not hidden and replaced after having been aborted?' 'For now I would be silently asleep and would rest in my dreams together with the rulers and princes of the earth, who have built isolated houses for themselves and have filled them with their silver.' The point at hand could not have been more clearly stated than in these words from the mouth of Job himself.[18]

As was said above, different spirits reside in different bodies, and their ranks are distinguished by a definite order and justice. Origen, Pythagoras and the Platonists list humans among the demons, including those who are not good but who could become good or evil as they live out their lives in a better or worse way. This is why both Christian theologians and the better philosophers say that life is like a road and a transition, a journey and a fight. The same judgement applies to other types of beings. Furthermore, we know that the best things into which a soul or spirit enters are the things which persevere the longest. That is what we said at the beginning: namely, all spiritual substances reduce to one, all material substances to three, there is one soul, one God, one first mind above all things and one soul of the universe.

Also, it is very probable that all illnesses are due to evil demons, which

[17] Mount Cicada (or Cicale or Cicala) is adjacent to Bruno's native town of Nola, immediately east of Naples. Its foreboding appearance made a deep and lasting impression on him during his childhood.

[18] See Job, 3: 3–20.

are expelled and replaced by their opposites with chants, prayers, meditations and ecstacies of the soul. And it cannot be denied that, in some people, there are dominating spirits who have the power to dispel certain types of illnesses. They say that Cyrus and other Persian kings could cure diseases of the spleen with the touch of their thumb. And it is well-known and clearly established that the same is true of the Kings of France, who cured disorders of the lymph glands with the touch of a thumb. It is also said that someone who is the seventh son of the same father, and who was born without the help of a woman, can do the same thing with his saliva.

One can prove that demons are material, and that they are of several different kinds, by the fact that they have emotions, desires, angers, jealousies and similar feelings found in humans, and in animals composed of observable and more dense matter. That is why the slaughtering and sacrifice of animals were instituted, for these demons are pleased a very great deal by such ceremonies and their fumes. It must be that these demons are constituted very much like us, because they also express their affections for some peoples and nations, while they detest and hate all others.

Some of these demons have names and are famous and more powerful, while others are more ordinary. The Romans called the latter 'gods of the dishes',[19] i.e., there were no specific offerings and sacrifices made to them. It is credible that such offerings were not necessary, but rather were pleasing to them (for they could provide for themselves whatever they needed). Nevertheless, these offerings were established for them as luxuries, which they would not have had without human contributions. For although they are able to know much more than we can, they cannot do and change as many things as we can, because of their spiritual and more noble and more reasonable characters. They are delighted by sweet scents and were adequately paid homage at one time by incense, saffron, moss, amber and fragrant flowers.

The more noble and more eminent spirits are said to be pleased by hymns, chants and musical instruments.

Above all, these are the gods who, by nature, 'have no need of us, and are neither favourably influenced by our merits, nor touched by our anger'.[20] Being affected by our good or evil actions pertains only to those spirits who can ask and receive from us some arrangement whereby they can have a better and happier life. This does not seem to be at all appropriate for those spirits who already enjoy a most happy state.

[19] 'petallares dei'. See Plautus, *Cistellaria*, II, 1, 46.

[20] Lucretius, *De rerum natura*, V, 650–1.

Finally, it must be consciously accepted and firmly asserted that all things are full of spirits, souls, divine power, and God or divinity, and that the whole of intelligence and the whole soul is everywhere, although they do not do everything everywhere. The poet has taken this idea from the teachings of Pythagoras.

> To begin: the heavens, the earth, the water wastes,
> the lucent globe of the moon, the sun, the stars,
> exist through inward spirit. Their total mass
> by mind is permeated: hence their motion.
> From mind and spirit comes life – of man, of beast,
> of bird, of monsters under the foam-flecked seas.[21]

The same message is contained in the sacred mysteries received by all people. Thus, in the Psalms and in the Book of Wisdom, it is said, 'The spirit of God has filled the whole earth and everything which it contains', and elsewhere, 'I fill the heavens and the earth'.[22]

A material substance differs from the substance of the mind and soul and sublime spirit as follows. The universal body is contained as a whole in the whole universe, but the spiritual substance is contained as a whole in each part. Thus, it exists everywhere as a whole and conveys everywhere an image of the whole, sometimes more clearly and sometimes more obscurely, sometimes in one way and sometimes in many ways. Thus, the entire nature of its form and light is reflected as a whole by all particles of matter, just as the universal body is reflected by all of matter.

This can clearly be seen in the case of a large mirror which reflects one image of one thing, but if it has been broken into a thousand pieces, each one of the pieces still reflects the whole image. Again, when different parts or bodies of water are separated from the whole or from the universal ocean by Amphitrite,[23] they have different names and properties; when they later flow together into one ocean, they have the same name and properties. Thus, if all the spirits and parts of air were to flow into one ocean, they would produce one soul, which elsewhere is innumerably multiplied. As a result, the philosophers say that in the original state of things there was one matter, one spirit, one light, one soul and one intellect.

[21] Virgil, *Aeneid*, VI, 724–9. Translation is by Frank O. Copley in *Virgil, The Aeneid* (Indianapolis: Bobbs-Merrill, 1965) 137.

[22] Wisdom, 1:7; Jeremiah, 23:24. Bruno's reference to the Psalms is apparently an error.

[23] Amphitrite was the goddess of the ocean and the wife of Poseidon. For Bruno, she represented the maternal origins of all things.

Let us now turn our attention to the many bonds between spirits. This is where the whole teaching of magic is to be found.

1. The first bond which ties spirits together is general in character and is represented metaphorically by the three-headed Cerberus of Trivia, the doorkeeper of hell. This is the triple power which is needed by one who binds, i.e., by the magician: namely, physics, mathematics and metaphysics. The first is the base; the second is the scale; the third is the summit of the scale. The first explains active and passive principles in general; the second explains times, places and numbers; the third explains universal principles and causes. This is a triple cord which is difficult to break.

2. The second bond is also triple and is needed in the agent, in the action and in the thing on which the agent acts. It consists of faith or credibility, of invocations, of love and of strong emotions in the application of the active to the passive. The role of the soul is to produce changes in the body of the composite, and the role of the body is to change the soul materially. If these bondings do not happen, or especially if they are not present, then no amount of attention or motion or agitation will produce any results. For a magician is most fortunate if many believe in him, and if he commands great persuasion.

3. The third bond, which is the source of effectiveness, is the number of the principles, which are distributed according to the four sectors of the universe and which are needed for actions which occur in the heavens and in nature. In addition, there are other principles needed for volitional and preternatural effects, but they do not have a specific location.

4. The fourth bond is the soul of the world, or the spirit of the universe, which connects and unites everything with everything else. As a result, everything has access to everything else, as was said above.

5. The fifth bond is the souls of the stars and the principles of places, of the winds and of the elements.

6. The souls of demons which preside over times, days, storms and the elements themselves.

7. The souls of men who are tyrants and rulers, and of those who have acquired some degree of fame and thus have become spirits.

8. The divine names and the names of the divine orders.

9. Markings and symbols.

10. Strong invocations and supplications to make the power of the superior overcome that of the inferior, for example, to banish evil demons by good ones, and to banish lower evil demons by higher ones. These demons are enticed by sacrifices and holocausts; they are frightened by threats, and they are summoned by the powers of inflowing rays of light.

11. By the power of the threefold world: elementary, celestial and intellectual.

12. The disposition to ask good things from good people, for example, chastity, honesty, purification and abstinence.

13. The adoption of cults and natural things in which there reside spirits which are similar to those required for actions.

14. The assessment of cults according to their different qualities.

15. The force of consecration which comes from perseverance, from prayer and from rituals.

16. A knowledge of feast days and of the days and hours of good and bad luck.

17. A knowledge of the different objects and methods found in religious observations in regard to the purity of their locations, and in regard to ablutions, contacts, endings, clothing, incensing and sacrifices.

18. The use of active and passive powers, for example, in the first or nearly first elements, and in stones, metals, plants and animals, in accordance with fourteen conditions.

19. Rings.

20. The techniques of enchantment.

In addition to these general bonds, others are listed in sixteen articles in the teachings of Albert.[24] Some of these are mentioned here, while others are not.

[24] Albert the Great (1193?–1280).

On the bondings of spirits, and first those arising from the three conditions of agent, matter, and application

For actions actually to occur in the world, three conditions are required: (1) an active power in the agent; (2) a passive power or disposition in a subject or patient, which is an aptitude in it not to resist or to render the action impossible (which reduces to one phrase, namely, the potency of matter); and (3) an appropriate application, which is subject to the circumstances of time, place and other conditions.

In the absence of these three conditions, all actions are, simply speaking, always blocked. For even if a flute player is perfect, he is blocked by a broken flute, and the application of the former to the latter is useless. Thus, a lack of power in the matter makes an agent impotent and an application unfitting. This is what was meant when we said that an absence of these three conditions, strictly speaking, always blocks an action.

Closer examination may show that the defect is due to only two, or even only one, of these conditions. But a defect in any one of them should be understood as meaning a defect in all three, as when the flute player and his performance are perfect but the flute is defective, or when the player and the flute are perfect but the performance is interrupted. If the whole meaning of efficient action is taken to consist in the application, then the first condition merges with the third, for the agent is nothing other than the applicator, and to do something is nothing other than to apply something.

Not all things are by nature passive, or active, in relation to all other things. Rather, as is said in the *Physics*,[25] every passion is from a contrary, and every action is on a contrary, or more specifically, on a disposed contrary, as is stated in the common saying, 'Active powers act on a properly disposed patient'. From this, it is clear that water mingles and mixes with water because of a similarity or awareness or sympathy, such that after they have united, no device can separate the one from the other.

Indeed, pure or unmingled wine also easily mixes with water, and *vice versa*, thus forming a mixture. But the parts of the wine contain some amount of heat and air and spirits, and thus the wine is not completely sympathetic with the water. As a result, they do not mix at the smallest level but survive separately to a noticeable degree in a heterogeneous compound, so that they can be separated again in various ways. The same thing

[25] Aristotle, *Physics*, I, 5. See also his *De generatione et corruptione*, I, 7, which makes this point more explicitly.

happens to sea water, which yields fresh water when it is distilled or filtered through wax containers. This would not happen if the mixture had been perfect. Furthermore, oil will never mix with water because the parts of oil cohere and are glued together like lovers, and they neither penetrate nor are penetrated by the parts of water. Therefore, anyone who studies the mixing of bodies with each other should give a great deal of attention to the condition of the parts, for not everything can be mixed with everything else.

Thus, one must study the arrangement, composition and differences of the parts, for a whole can be penetrated by a whole in one direction but not in another. This happens in all things, like stones, wood and even flesh, which are penetrable, or more penetrable from one side or direction than another. This is clear when fluids are expelled by pushing along the length of fibres. And wood is more easily split lengthwise, for wood is more easily penetrated along its length than its width because the pores located between the fibres create tubes or passages in that direction.

Furthermore, one must not only examine the character and arrangement of the parts, but also the condition of the whole structure, for certain passions are naturally adapted to be received by one subject rather than by another. For example, a torpedo fish causes a shock to the hand of the fisherman, but not to the net. And, as the old joke says, the fires of love burn the heart and the breast, but leave the chest cold and uncooked.

The same thing happens with thunderbolts, which have at times liquified a steel sword without damaging its scabbard. An astonishing event also happened in Naples to a very beautiful and noble young girl whose pubic hair was burned, but nothing else. They also say that when the wood of a barrel was burned away, the wine remained firm and solid without it. Many such things have happened because of this ultimate occult power which resides in the atoms of this kind of fire and which acts in one place but not in another. The laurel and the eagle are used as insignia by generals and poets because they are never touched by lightning, and so like them, generals and poets are friends of Apollo and Jupiter.

What happened to that young girl does not happen to just any human being. The reason for this is that not all people have the same physical constitution and temperament and the same quality of spirit, and, as a result, not all have a soul that can stop the rains and command the winds and the storms. The astonishing things that happen in bodies must be related to a special constitution which, because of the innumerable differences in them, is due sometimes to the laws of the whole species and sometimes to a special prerogative of the individual.

Hence, magicians carefully examine both species and individuals in order to grasp the effects of their power. Being prudent leaders, they recruit as their soldiers and gather as their military aides not those who are friends or well-known or highly recommended people, but rather those who are more favoured by fortune and those who usually are lucky enough to avoid such dangers. Likewise, by wearing and carrying and otherwise using certain plants and minerals, they try, as if by means of direct contact, to appropriate for themselves certain prerogatives of power. And thus, as leaders protected by laurel crowns, they do not fear the lightning.

Next it should be noted that for specific animals there are poisons, like hemlock generally is for humans, which usually are consumed as a very helpful nourishment by other animals, and which readily fatten them. Likewise, it should be seen that for various species there are different foods, poisons and antidotes. An important principle of magic and of medicine is to be able to distinguish the different constitutions and explanations of illnesses and good health, and the principles of changing or preserving their forms and dispositions by using external objects. Thus, the chemist knows that nitric acid acts on hard things, such as iron, silver and bronze, but not on gold or lead. He also knows that quicksilver absorbs oil very rapidly, but gold completely rejects and repudiates it. Furthermore, the seeds or juice of verbena plants are strong enough to break up stones in the bladder yet seem to do no damage to flesh, bones, membranes and other parts of the body.

There are those who explain these facts in terms of the pores being wide or narrow. I would readily grant this in some cases, but not in most cases, nor in the more important cases which are discussed above, for the reason why nitric acid penetrates one thing more than another is not that the one has wider pores. Likewise, the spirit of the verbena plant breaks up stones but not bones and flesh, even though the latter has larger pores. And what would they say about diamonds which are not split by fire, the smallest and most penetrating of bodies, even though they do absorb the blood of a billy goat?

Therefore, one must maintain the following general principle: not all things are influenced by everything else, and not all effects happen to everything in the same way. To give a proper explanation, the reason must be found in individual effects and cases. The occult forms and differences in things do not have their own names. They are not observable by means of vision and touch, and explanations of their specific origins are not to be found in visual and tactile differences. All we can say about these occult forms is that they do exist. As a result, we conclude that not even the demons themselves could talk about them easily, if they were to choose to discuss them with us, using our words and the meanings which are signified by our words.

Secondly, the bondings arising from sounds and songs

A second type of bonding is based on the conformity between numbers, between measures and between times. This is the origin of those rhythms and songs which have such a very great power. Some people are affected more by tragedies, others by comic melodies, and others are affected generally in all cases. Some even react like that barbarian general who, when he heard musical instruments played very skilfully, said that he preferred to hear the neighing of his horse. He clearly proved by this that he was a disgrace and was unworthy of appearing to be human.

By the term 'songs' we intend to refer to much more than harmonies, for as some have experienced, the most powerful songs and poems seem to contain more discord than harmony. Perhaps such was the condition of the soul of that subhuman general who was more easily influenced by the sounds of his horse's neighing. For just like someone who looks at the sensible harmonies of vision, the souls of humans and horses and dogs are captured by different harmonious sounds, and different things are beautiful according to the condition of each species. As is said in the proverb 'from an ass to a lyre', not all songs are well suited for everyone. And as different harmonies bind different souls, so also different magicians bind different spirits.

These bonds are tenacious for two reasons. First, they are perceived or encountered in the soul through hearing, just as the voices of the Marsi and the Psylli became such powerful voices when they were present in the serpent.[26] Second, the bonding effect is brought to completion by an occult murmur which, analogously to the relations between spirits, did not originally come from the binder to the bound for the purposes of bonding. For those who are enchanted do not always hear the voice of their enchanters, and even when they do, they are not sensibly affected immediately.

[26] Both the Marsi, a people who lived near Lago di Celano in Southern Italy, and the Psylli, a people who lived near Sidra on the North African coast, were well-known as snake charmers.

Further, one should note that the rhythm or characteristics of one sound can mingle with, and obscure, the rhythm of another sound. As a result, it is said that when a wolf, or some say a deer, is seen by people who are bonded to the spirit of that animal, they lose their voices and cannot easily form words. And they say that when a drum made of sheep skin is located next to a drum made of wolf skin, the former loses its sound, even though otherwise it emits strong sounds when forcefully hit. The reason for this is that the spirit which somehow remains in the dead wolf skin can bond with, and control, the spirit in the sheep skin, and thus they are subject to the same antagonism and dominance which are present in the living animals.

I have not personally experienced this. But it is a possibility and is reasonable, even though this relationship is not always found between living things and between species. The ass fears the wolf no less than does the sheep, and is equally frightened by its danger. Nevertheless, a drum made of the skin of a wolf does not diminish the equally strong beats from a drum made of the skin of an ass, but rather increases their loudness considerably.

Let us next consider lyres whose strings are made of the tendons of sheep and wolves, which are always opposed. It is well-known by many that if two lyres or cithers are constructed in the same way, and if a string of only one of them is plucked, its sound is not only consonant with the string of the other, but it will generate the same motion in the other. This, indeed, is quite understandable. It also happens that, through a certain sound or gesture or other such thing, the presence of one person affects the soul of another person, and thus an indissoluble friendship arises. There are those we dislike without reason as soon as we see them, and also others we love without cause. This love and hate are sometimes reciprocal and sometimes not. This happens because of the domination of the one person over the other in respect to one type of feeling, which in turn is blocked by another type of feeling in the other person. Thus, we are attracted by a feeling of love for one type of dog or bird, while they may be struck by fear, and thus avoid and dislike us.

This type of bonding also includes prayers and petitions, which some use to solicit both peers and superiors in cases where considerations of justice, honesty or reason produce no results. Sometimes proposals from fools and buffoons are so effective that people who are clever try to ensnare the souls of their superiors by playing such a role rather than by using more proper means. This happened during the papacy of Julius III, who rejected and dismissed those who would pray, beg or cry. But if someone approached him with humour and wit, after kissing his foot, that person would be able to get from him whatever he wanted.

We might also consider the art of speaking and its type of spiritual bonding. This occurs in songs and poems and in whatever orators do to persuade, to dissuade and to move the emotions. The orators omit the other parts of this art and try to hide them in the lap of magicians or philosophers or those versed in politics. But Aristotle has covered most of it in his *Rhetorica ad Alexandrum*,[27] where he organizes his considerations under two headings. He examines first what the speaker needs and finds helpful, and second what is pleasing and amusing in what he says or binds, by considering his habits, status, conclusions and practices. But this is not the place to recall and review all these matters.

Thirdly, the bondings arising from vision

The spirit is also bonded through vision, as has been said frequently above, when various forms are observed by the eyes. As a result, active and passive items of interest pass out from the eyes and enter into the eyes. As the adage says, 'I do not know whose eyes make lambs tender for me'.[28]

Beautiful sights arouse feelings of love, and contrary sights bring feelings of disgrace and hate. And the emotions of the soul and spirit bring something additional to the body itself, which exists under the control of the soul and the direction of the spirit. There are also other types of feelings which come through the eyes and immediately affect the body for some reason: sad expressions in other people make us sad and compassionate and sorry for obvious reasons.

There are also worse impressions which enter the soul and the body, but it is not evident how this happens and we are unable to judge the issue. Nevertheless, they act very powerfully through various things which are in us, that is, through a multitude of spirits and souls. Although one soul lives in the whole body, and all the body's members are controlled by one soul, still the whole body and the whole soul and the parts of the universe are vivified by a certain total spirit.

Hence, the explanation of many spiritual feelings must be found in something else which lives and is conscious in us, and which is affected and

[27] In 1587, Bruno wrote a commentary on this Aristotelian treatise under the title *Explicatio rhetoricae Aristotelis ad Alexandrum*.

[28] Virgil, *Eclogues*, III, 108.

disturbed by things which do not affect or disturb us. And sometimes we are touched and injured more significantly by those things whose assaults we are not aware of than we are by things which we do perceive. As a result, many things which are seen, and forms which are absorbed through the eyes, do not arouse any consciousness in our direct and external sensory powers. Nevertheless, they do penetrate more deeply and lethally, so that the internal spirit is immediately conscious of them, as if it were a foreign sense or living thing. Thus, it would not be easy to refute some of the Platonists and all of the Pythagoreans, who believe that one human person of himself lives in many animals, and when one of these animals dies, even the most important one, the others survive for a long time.

Hence, it would obviously be stupid to think that we are affected and injured only by those visible forms which generate clear awareness in the senses and the soul. That would not be much different from someone who thinks that he is injured more or less only by blows of which he is more or less conscious. However, we experience more discomfort and suffering by being pricked by a needle or by a thorn irritating the skin than we do by a sword thrust through from one side of the body to the other, whose effect is later felt a great deal more, but at the time we are unaware of the injury caused by its penetration of parts of the body.

So, indeed, there are many things which stealthily pass through the eyes and capture and continuously intrude upon the spirit up to the point of the death of the soul, even though they do not cause as much awareness as do less significant things. For example, seeing certain gestures or emotions or actions can move us to tears. And the souls of some faint at the sight of the spilling of another's blood or in observing the dissection of a cadaver. There is no other cause of this than a feeling which binds through vision.

Fourthly, the bondings arising from imagination

The role of the imagination is to receive images derived from the senses and to preserve, combine and divide them. This happens in two ways. First, it occurs by the free creative choice of the person who imagines, for example, poets, painters, story writers and all who combine images in some organized way. Second, it occurs without such deliberate choice. The latter also happens in two ways: either through some other cause which chooses and selects, or through an external agent. The latter, again, is twofold. Sometimes the agent is mediated, as when a man uses sounds or appearances to bring about stimulations through the eyes or ears. And sometimes the agent is unmediated, as when a spirit, rational soul or demon acts on the imagination of someone, asleep or awake, to produce internal images in such a way that something seems to have been apprehended by the external senses.

Consequently, some possessed people seem to see certain sights and hear certain sounds and words which they truly think are caused by external subjects. Hence, they strongly and persistently assert that what they have seen and heard is true, when it is their reason which is deceived, and not their senses, for they do hear what they hear, and they do see what they see. But the very same thing which they think they see as derived from external sounds in their ears and from external sights absorbed by their vision are fantasy images presented to their internal sense. However, they think that these impressions of the internal sense are the real things. Thus, it happens that they refuse to be recalled to a healthier point of view by actual witnesses, whom they prefer to reject in favour of their own imagination, and who they truly think are deaf and dumb. Medically these matters are cases of mania and melancholy, and are called 'the dreams of the wakeful'.

Further, this type of bonding is not due simply to a material principle, as is believed by certain well-known medical people with an obstinacy which is most crude and oppressive. Nor is it due simply to demonic or diabolical efficient causes, as is believed on their part by some theologians. Rather, both causes co-operate. The material factor is a melancholic humour, which we call the kitchen or the bathhouse of the saturnalian demons. But the efficient cause and moving spirit is a demon who does not have a completely immaterial substance, because these demons seem to be endowed with many animal affections and have definite properties of denseness. Although they are spiritual substances, nature has given them a body which is very thin and is not endowed with senses. They belong to that genus of animal which, as was said, has more species than do living, composite and sensory animals.

Now, a specific soul comes to a specific seed which has been properly deposited in a specific place, or conversely, a body makes or produces from itself, as it were, a specific animal form or living thing. For example, from one seed the olive is born, from another a dog, from another a human, and in general one thing or another is suited to be born from a body which is structured in one way or another. As the poet says, 'The eggs come more readily to where the seed is sown'.[29] As a result, like the proper seed being sown in the proper field, good and evil spirits and the beginnings of consciousness are born from a proper mixture and combination of specific hearts or brains or animal spirits, and conversely, improper mixtures produce disturbances. These results are mutual: certain souls bring certain bodies into existence, and certain bodies bring certain souls into existence, in accordance with what are called the substantial and the specific difference and subsistence.

When two spirits approach and come near to each other, either because of an accidental combination or because of objects attached to the body, then the dominance of a raging spirit can be removed safely and methodically. This is done either by incantations, that is, by rhetorical and friendly and curing persuasions which restore the besieged spirit; or by the expulsion and evacuation of noxious material with purgative medicines; or by foods and a happy, sunny atmosphere which are agreeable to human life, and which introduce better matter for the spirit; or by soothing and moderating the harmful materials which sometimes enter into the mix.

As a result, the spirit alone does not produce these living animal operations, nor does the body do this without the spirit. Rather, for these things to occur, whether they be good or bad, or in accord with or in opposition to the nature of the species, what is required is both a material principle and a formal or efficient cause of the needed type. Further, it is reasonable to say that a simple purgation of humours and a simple diet are adequate to cure disturbed images and to free the internal senses which are bound in this way.

However, from this, one cannot accept the conclusion drawn by a most stupid and dull-witted medical man in his book *De occultis naturae miraculis*[30] (*On the Hidden Miracles of Nature*), which presents more nonsense than words and sentences can describe. He concludes that spirits are the same thing as humours because the expulsion and evacuation of humours also expels and evacuates these spirits with their marvellously independent and structured powers. In this way, with equal justice, one could say that, because the excellence of the soul forces it to leave the body and be many souls in succession, he should think that the soul, itself, is a humour or excrement. Or if he himself were to decide to abandon his house and country because a shortage of food and water made him ignorant of medicine and of the obvious colours and sounds of nature, we should conclude that

[29] Virgil, *Georgics*, I, 54. [30] Levinus Lemnius, *De miraculis occultis naturae* (Anterpiae, 1559).

he himself belongs to the same species as the things which expelled him.

Since the senses happen to be bound and obligated in all these ways, magic and medicine must pay very special attention to the workings of the imagination. For this is the doorway and entrance for all the actions and passions and feelings of animals. And to that linkage is tied the more profound power of thought.

Fifthly, the bondings arising from thought

The bondings of the imagination would not be very significant in themselves if they did not duplicate the powers of thought, for those appearances which bind and obligate the souls of those who are simple-minded, stupid, credulous and superstitious, are derided and condemned as empty shadows by those who have a sober, disciplined and well-bred mind. As a result, all practitioners of magic, medicine and prophesy produce no results without a pre-given faith,[31] and unless they act according to the rules of that faith. (We use the word 'faith' here in the more general sense in which it is used by these people, individually and as a group.)

This faith arises in some people from their pre-given powers, which are well disposed and organized, and in others, it comes from a disturbance of their powers. Indeed, great results are produced by those bonds which come from the words of a man of eloquence, by which a certain disposition arises and flourishes in the imagination, which is the only entrance for all internal feelings and is the bond of bonds. This is the point of Hippocrates' saying, 'The most effective doctor is the one whom most believe'. The reason for this is that he binds many people with his eloquence or presence or fame. This applies not only to medicine but to any type of magic or to any power identified by a different title, for, in the act of binding, the imagination must be stimulated or else one can hardly motivate anyone by other means.

In regard to the notion that it is possible for a person to do everything on his own, the theologians believe, agree and state publicly that it is impossible to help those who do not believe the minister. The reason for their lack of power lies in the imagination which they cannot bind. Indeed, kinsmen reject and laugh at physicians and divines because they know about their humble origins and education. As the well-known adage states, 'No one is a prophet in his own land'.

[31] In his *De magia mathematica*, 8, Bruno claims that even Christ could not cause miracles when the disciples around him had too little faith. See Matthew, 17:19.

Thus, someone who is less well-known can bind people more easily. Given a good general impression and a disposition to be believed, he can somehow use the power of his soul to arrange, disclose and explain things for them, as if windows which had been closed are opened to receive the light of the sun. This opens the door to those other impressions which the art of binding seeks in order to establish further bonds, namely, hope, compassion, fear, love, hate, indignation, anger, joy, patience, disdain for life, for death, for fate, and all of the powers which cross over from the soul to change the body.

There is no need for a more detailed investigation and consideration of the changes which occur to the types of bondings which follow upon faith and a good impression, and which were just listed above. Further, it is not our business at present to examine the more spiritual powers of the soul which follow next: namely, memory, reason, experience, intellect and mind, because the acts of these powers do not flow over into the body and change it. Rather, all physical changes originate from the powers which are prior to thought and which are its principal and efficient causes.

As a result, all magical powers, active and passive, and their species are dependent upon magical bondings. As Plotinus has asserted, both the wise man and the fool can be bound by the natural principles residing in them, unless the subject also contains some principle which can reject and dismiss magical influences. For as was said above, not everything enters into everything else, and not everything mixes with everything else, as, for example, water and oil do not mix. As Plotinus himself has stated, and as Porphyry confirms in his *Vita Plotini* [*Life of Plotinus*], the evil spells with which a certain Egyptian tried to bind and injure Plotinus were turned back against him.[32] These things are discussed in our *De vinculis in genere* [*A General Account of Bonding*].

32 The incident related here can be found in Porphyry's *On the Life of Plotinus and the Arrangement of his Work*, §10, in Plotinus, *The Enneads*, translated by Stephen MacKenna, Third Edition (London: Faber and Faber Ltd., 1962) 8.

A General Account of Bonding

A general account of bonding

Anyone who has the power to bind must to some degree have a universal theory of things in order to be able to bind humans (who are, indeed, the culmination of all things). As we have said elsewhere, in this highest species, it is possible to see, and especially to rank, the species of all things. For example, some humans are like fish, others like birds, others like snakes, and still others like reptiles, whether it be in the latters' species or in their genera. Also, different people have different functions, habits, purposes, inclinations, understandings and eras. And so, as was imagined by Proteus and Achelous, the same material object can be changed into different forms and figures, such that to bind them continuously one should always use differing kinds of knots. In addition to this, let us notice the conditions of human life: being young and then old; being of a moderate station, or noble, or rich, or powerful, or happy, or, indeed, even envious and ambitious; or being a soldier or a merchant, or one of the many other officials who play a role in different ways in the administration of a state, and thus who must be bonded to each other because they function as agents and instruments of the state. In effect, it seems that nothing can fall outside of an examination of civil life when it is considered in this way (whether it be bonding, or being bonded, or the bonds themselves, or their circumstances). This is the reason why we have assembled the following considerations, which are entitled *A general account of bonding*.

On bonding agents in general

1. *Types of bonding agents.* Taken universally, bonding agents are God, demons, souls, animals, nature, chance, luck and, finally, fate. This universal

force of bonding, which cannot be designated by one name, does not bind because of the nature or the sensitivity of matter. A body does not have any feeling on its own, but only because of a certain force which resides in it and which emanates from it. This force is called, metaphorically, the 'hand which binds', and it is oriented and adapted to bonding in multiple ways.

2. *Effects of the bonding agent.* As the Platonists say, it is this bonding agency which adorns the mind with orderly ideas; which fills the soul with sequential arguments and harmonious discourse; which makes nature fertile for various seeds; which structures matter in innumerable ways; which vivifies, soothes, caresses and activates all things; which orders, generates, rules, attracts and inflames all things; and which moves, reveals, illuminates, purifies, pleases and completes all things.

3. *How art binds.* An artisan binds with his art, for art is the excellence of the artisan. Even someone who is stupid and dull witted will see the beauty of natural and artificial things, even though he cannot at the same time grasp and admire the talent which has generated all things. For him, 'the stars do not speak of the glory of God'.[1] Rather, like a brute animal, he will shower his affections not on God but on His effects.

4. *Humans are bound in many ways.* Of all the things which bind, certainly more of them bind humans than brute animals, and more of them bind those who have an active character than those who are dull witted; those who are well endowed in their faculties and powers are aware of more details, circumstances and purposes, and thus, they are moved by more desires.

5. *How the senses are panderers for the bonding agent.* Dull witted people are bound by lusts, which are aroused infrequently and by natural impulses, and which are few in number and limited to base nourishments. Such people are not soothed by eloquent speech, nor are they won over by beauty, music, painting or by any of the other attractions of nature.

6. *Why only one bond is not enough.* As I am bound by more things, I become aware of the many things which bind me, for there are many different kinds of beauty. Thus, I am inflamed and bound in a relationship by one thing in one way and by other things in other ways. If every relationship were reduced to one, then perhaps one thing would be welcomed for all purposes

[1] This is the first line of Psalm 18, to which Bruno has added the negative.

and for all occasions. But up to now, this has not happened in nature, which has spread about many bonds of beauty, happiness, goodness, and the various contraries of these dispositions, and which widely distributes them separately according to the numerous types of matter. But it does sometimes happen that a person is so tied to one object that his awareness of other things is weakened, overwhelmed and suppressed, either because of the dullness of the senses which are blind to and neglectful of all other things, or because one bond is so strong that it weakens and distorts him. But this is extraordinary and happens rarely and in only a few cases. For example, there are some whose souls seem to be so carried away by the hope of eternal life and by a vivid faith and credulousness, and seem to be so separated from the body in some way, and so strongly bound and controlled by some object in their fantasies and in their opinions, that they do not seem to be aware of the most horrible torments. This clearly happened to the philosopher Anaxarchus, to Andrew[2] the Galilean, to the priest Lawrence,[3] and to others up to our own day, who were murdered by rulers and kings for the sake of their religion. This also happened for the sake of reason to Diogenes the Cynic and to Epicurus, who argued that they could banish all awareness of pain and pleasure by binding their souls, according to natural laws and principles, with a contempt of all things and of every type of opinion ... They thought they would attain the highest good available in this life to the human species by preserving their souls in a state of heroic pleasure above sorrow, fear, anger and other feelings. They claimed that, by holding in contempt the ignoble things in this very transitory life, they could attain a life similar to the gods even while in this mortal body. They thought that they had actually attained this highest good and sublime virtue, and that they had shown this to others.

7. *What power contributes to a bonding agent.* There are those who say that a bonding agent of greater power binds something else which in turn does not bind it; if the powers are equal, then there is a reciprocal bond which consists in a balance of that quality. But it would follow from this opinion that bonding powers are continually changed and altered as forms, circumstances and natures are altered, for a young man does not bind the same things which he bound as a boy, and a woman does not bind the same things which she bound as a girl. Hence, a bonding power is not simple or reducible to only one thing, but is composite, variable in nature and composed of contraries.

[2] St Andrew the Apostle, brother of Simon Peter. [3] St Laurentius, third-century Christian martyr.

8. *What is bound more easily*. A person who is more truly human is bound most strongly by the most worthy things, and he prefers much more to seek out more worthy things than to possess base things, for certainly, we are easily irritated by base things and more ardently seek for things which we do not easily attain.

9. *That the same thing bonds contraries in the same way*. Bonding agents which pertain to the same type of bonding seem to be confusing, and in a sense even contradictory, when one considers the contrasting effects and circumstances of the bonds. Consider, for example, the bonds of physical love, which seem to be both a fire and snare at the same time, which drive one to shout and to be silent, to joy and to sorrow, to hope and to desperation, to fear and to boldness, to anger and to gentleness, to weeping and to laughter. Hence the verses:

> I, who carry high the standard of love,
> have frozen hopes and burning desires:
> at the same time I tremble and freeze, burn and spark,
> I am mute and fill heaven with ardent cries.
>
> From the heart I sparkle and from my eyes I shed water;
> I live and I die, laugh and lament.
> The waters live and the fire does not die,
> for in my eyes I have Thetis and in my heart, Vulcan.[4]

10. *A bonding agent does not bind different things with the same bond*. A thing is not absolutely beautiful if it binds only playfully; it is not absolutely good if it binds only usefully; it is not absolutely large if it is limited. Regarding beauty, notice how monkeys and horses please each other; indeed, not even Venus pleases some types of humans and heroes. Regarding goodness, notice how all things contain contraries, and how different animals find what is good for them under the seas or on dry land, in mountains or in fields, in abysses or on summits.

11. *He who binds*. Therefore, he who knows how to bind needs to have an understanding of all things, or at least of the nature, inclination, habits, uses and purposes of the particular things that he is to bind.

[4] The translation of these two quatrains is taken from John Charles Nelson, *Renaissance Theory of Love* (New York: Columbia University Press, 1958) 222. Bruno quotes these verses, with slight variations, from his *Eroici furori*, Part 1, Dialogue 2, first paragraph.

12. *No one particular thing can bind everything.* What is absolutely beautiful and good and large and true binds every feeling and every mind absolutely. It destroys nothing; it contains and seeks out all things; it is desired and pursued by many because it invigorates with different types of bonds. Hence, we abundantly acquire many skills, not to be able to act universally and simply, but rather to do this at one time, and that at another time. Thus, since no particular thing is absolutely beautiful, good, true, etc., whether it be above all genera or in any particular genus or species, it follows that nothing can bind simply at any of these levels. Nevertheless, there is a desire for the beautiful, good, etc., in all things, for everything seeks to exist and to be beautiful in every way, at least according as its species and genus allow. Beauty and goodness are one thing for one species, and another thing for another; in one thing one contrary dominates, and in another the other dominates. The total beauty and goodness of one species cannot be attained except through the whole species for all eternity and in each of its individual members taken separately. Testimony to this in regard to human beauty is given by Zeuxis in his painting of Helena, whom he selected from among the young women of Crotona. Although he has given us a girl who is beautiful as a whole and in every detail, how could he have ever presented complete beauty in every way, since the different types of physical beauty in the female species are innumerable, and only some of them can be found in any one subject? For beauty, which consists of a special symmetry or of some other incorporeal aspect of physical nature, occurs in a myriad of forms and arises from innumerable ordered patterns. Thus, just as the rough surface of a stone does not meet, fit and adhere to the rough surface of any other stone, except when their folds and cavities correspond a great deal, likewise not every quality will reside in any soul. Therefore, different individuals are bonded by different objects. And even though the same object bonds both Socrates and Plato, it binds each of them in a different way. Some things excite the masses, other things affect only a few; some things affect the male and the manly, other things the female and the feminine.

13. *The various instruments of the bonding agent.* Nature has distinguished, dispersed and disseminated the objects of beauty, goodness, truth and value in its own way. And, as a result, different things can bind for various reasons and for different purposes. For example, a good farmer binds and becomes admirable for one reason, a cook for another reason, a soldier for another, and a musician, a painter, a philosopher, a boy, a girl, for different

reasons; one of them walks better, while another speaks better. No one of these alone possesses all things in all ways. Rather, the one who is found to be happy and skilful in more ways and at more levels will bind more things, will rule in more ways and will win out over more people of their own species.

14. *The opportunities for the bonding agent.* We experience various feelings at different times and on different occasions, and there is no one measure common to them all. Likewise, there is no one and simple factor which can please everyone or satisfy all things, much less does any one thing satisfy different persons or one person at different times. For example, neither the same food nor the same quantity or quality of food always satisfies. This principle applies to all things which bond our appetites.

15. *The different types of bonding agents.* Some things bond by their own power. Other things bond because of their quantity or because of one of their parts. Still other things bond because they are aided by something else to which they are attached or which properly disposes them, as when a beautiful building arises out of irregularly shaped parts.

16. *The variable power of the bonding agent.* There are many things which we judge to be beautiful but which nevertheless bind us as good, for example, a horse, a ship, a house, a statue, a dog or a bird. But a beautiful person does not bind us in order to be considered good, and a good person does not bind us in order to be thought to be beautiful. It could happen that crime and error are joined to the beautiful. Consider a beautiful but poor woman: the more disturbing she is, the more easily one tends to give her a gift. There are diverse reasons for diverse things, contrary reasons for contrary things, and similar reasons for similar things.

17. *Where the bonding agent is located.* Those who have not studied the matter too deeply, like the Platonists, think that that which binds is the form of the thing, and crosses over from the thing to the mind, even though it does not leave the object itself. This is like fire which does not lessen when it communicates its form, and like an image which is in an object first, then in the mirror, then in the intervening space, and finally in the eye. But considering the matter more profoundly, we find that, indeed, it does exist in the body, and it consists of a certain physical bond, but, like the soul whose powers reveal its condition, it does not occupy any specific part of the body. Indeed, even though the amorous effects of love may arise from the eyes or

the mouth or the complexion, nevertheless it is clear that it is not found in them alone, nor does it arise from them alone. For the eyes, considered separately and by themselves, do not have the same force when they are not united with the other parts of the face. The same is true of the mouth, the nose and the complexion, which are not beautiful when depicted separately by a painter. As a result, the nature of beauty is indefinite and quite indescribable, and the same is true of the nature of goodness and cheerfulness. The complete nature of a bond is to be found not just in the object itself, but also in another equally important place, i.e., in the one who is bound. Whether food is greedily consumed or is returned uneaten after a meal, this makes no difference at all as far as the substance and quality of the food is concerned. And the bonds of love, which were intense before sexual intercourse, become relaxed when the seed is ejaculated and the fire becomes moderated, even though the beautiful object remains the same. Therefore, the whole nature of a bond cannot be found in the object.

18. *The predispositions of the bonding agent.* The bonding agent is said to be predisposed to bonding in three ways: by its order; by its measure; and by its type. The order is the interrelation of its parts; the measure is its quantity; and its type is designated by its shapes, its outlines and its colours. For example, in a bonding of sounds the order consists of a rising and falling through high, low and intermediate notes; the measure is the use of thirds, fourths, fifths, sixths, etc., and the progression of tones and semitones; the type is the harmony, softness and clarity. In all things which are predisposed to bonding, whether they be simple or composite, all three of these factors are present in a proportional way.

19. *The diversity of predispositions.* Regarding the bonds themselves, there is another predisposition: the signs and vestiges which reveal how well developed the soul is. These move the soul to seek out only the enjoyment of another soul, to which it becomes attached and united. And also because of this predisposition, when there is a suitable ordering of the body and of its parts and of the garments which clothe it, the soul is then bound to reach out for the enjoyment of the body. When this predisposition is present in both the soul and the body, it impels each more strongly to the enjoyment of the other, or each is attracted by the other principle. Furthermore, there are some people who are so bound by the soul that they also desire the very same body which contains the soul. There are even a few people who are so focused on the soul that they look down upon certain features of the

body unless they are predisposed by the soul. Thus, the famous story is told about Socrates, who required that an attractive young boy first speak out before he would declare his love for him.

20. *The condition of the bonding agent.* Flatterers give high praise to ordinary virtues, downplay faults, excuse errors, transform evil deeds into virtues, and act very cautiously so as not to reveal their art of flattery. As a result, they bind to themselves people who are not very clever, for to be loved and honoured is the most pleasing and delightful thing for anyone, and to be able to bind someone requires a certain higher type of virtue.

21. *How the bonding agent is bound.* He who binds experiences joy and a certain sense of glory, and this is greater and stronger insofar as the one who is bound is more noble, more worthy and more excellent. The strength of the bond by which he who binds is himself bound by the one who is bound is located in this sense of joy and glory. In praising the vanquished, the victors extol their own victory, and sometimes they even deceive themselves more than others, both in their desires and in the other public consequences of bonding. On the other hand, someone must be of a worthless character if they are so unpleasant as not to reciprocate in spirit with someone who loves them, when that person is honourable and distinguished, or with someone who is bound in spirit to them in some other way.

22. *The types of bonding agents.* There is one type of bonding in which we wish to become worthy, beautiful and good; there is another type in which we desire to take command of what is good, beautiful and worthy. The first type of bonding derives from an object which we lack, the second, from an object which we already have. These two types bind both what is good and what is thought to be good, although this bond always occurs in some proportional or suitable way. Also, fantasy and opinion bind more things than does reason, for the former are indeed stronger than the latter. To be sure, there are many who love without a reason, although there is some cause which motivates their love, and, as a result, they are bound but do not know what binds them.

23. *The blindness of the bonding agent.* The explanation of bonds is, for the most part, hidden, even from the wise, for what use is it to appeal to analogies, similarities, family traits and other such meaningless words when we see a person who hates nothing more than another person who is his genial companion, while at the same time and without reason, he also loves that

person more than anything else? A general explanation is useless in a case like this, because such an account does not distinguish between things which belong to the same genus or species, for example, between female and female, or male and female, as well as between other human conditions like being old or young. And what would you say about the love of things known only by hearsay, which is usually called 'devotion'? Are not humans bonded to higher and immaterial things, as well as to imaginary things, and especially to things beyond experience? I will pass over here any discussion of the specific types of binding powers, and especially of the power to bind through incantations. It is not true, as some have said, that the power of bonding is derived from what is good rather than from an opinion about what is good; nor is it derived from a known rather than from a hidden cause. We have already spoken above about the different types and species of goods.

24. *The diligence of the bonding agent.* Just as dull people are bound more easily by a shrewd flatterer than by a true friend, likewise, bonds and bonding powers are established and maintained in skillful ways. For example, a timid man recommends against joining the army; a strongly godless person recommends against becoming a priest; a cruel person recommends against caring for others. Things move more easily towards that to which they are inclined, just as someone who wishes to pick up something cylindrical attends to surfaces which are round rather than flat or angular.

25. *The weapons of the bonding agent.* The bonding agent has three types of tools. The first type is located within him and is two-fold: those which are essential or natural, i.e., those which belong to the nature of his species; and those which are accidental and acquired, which follow from the nature of his species, for example, sagacity, wisdom and art. The second type is located in his environment, for example, chance, good fortune, opportunity, encounters and arranged meetings. The third type is located above him, for example, fate, nature and the favour of the gods.

26. *The vicissitudes of the bonding agent.* The kind of proportionality which we regularly experience in eating and in sexual intercourse is found in every act of bonding. For we are not attracted and bonded by these desires and loves at all times, or in the same way, or in the same degree, or with the same variations of time. The reason is that our physical constitution and all of its consequences fluctuate and change with time. Therefore, the moment for bonding must be predicted ahead of time, with careful and

antecedent deliberation, and the opportunity must be quickly seized when it presents itself, such that he who can bind will act and bind as soon as possible.

27. *The eyes of the bonding agent.* Bonds are so subtle, and that which is bound is so barely sensible in its depths, that it is possible to examine them only fleetingly and superficially. They change from moment to moment and are related to the bonding agent like Thetis fleeing from the embraces of Peleus. It is necessary to study the sequence of the changes and how the power of a subsequent form is influenced by its predecessor, for although matter is indeterminate in relation to innumerable forms, still its present form is not equally distant from all the others. Rather, only one of those forms is the immediate successor, others follow after many or a few intermediaries, and one is located the furthest away of all. Thus, just as the form of blood immediately follows the form of chyle, so do the bonds of anger follow the bonds of indignation, and the bonds of sadness follow the bonds of anger, as yellow bile easily becomes dark. Hence, after having carefully observed the disposition and the present qualities influencing Thetis, Peleus planned and prepared ahead of time the bond to win her over before she might change into some other form, knowing full well that a snake and a lion and a wild boar are captured in different ways.

28. *The enticements of the bonding agent.* A bonding agent does not easily bind someone who can be bound, just as a military commander does not easily capture a well-protected fortress unless entry is provided by an internal traitor, or by some arrangement with a collaborator, or by surrender or by some sort of a compliant official. Thus, in her own realm, Venus does not bind and does not easily capture the fortress when goblets are empty, when the spirit is disturbed and when anxiety is aflame. But the fortress is handed over when the goblets overflow, and the soul is at rest, the mind is quiet and the body is at leisure. Having closely observed the changes of these guards and custodians, one must suddenly act with boldness, attack with force, use all resources and never hesitate. This same course of action must be followed in other acts of binding.

29. *The steps in bonding.* A bonding agent does not unite a soul to himself unless he has captured it; it is not captured unless it has been bound; he does not bind it unless he has joined himself to it; he is not joined to it unless he has approached it; he has not approached it unless he has moved;

he does not move unless he is attracted; he is not attracted until after he has been inclined towards or turned away; he is not inclined towards unless he desires or wants; he does not desire unless he knows; he does not know unless the object contained in a species or an image is presented to the eyes or to the ears or to the gaze of an internal sense. Bonds are brought to completion by knowledge in general, and they are woven together by feelings in general. I say 'knowledge in general' because it is sometimes not known which sense has captured the object, and I say 'feelings in general' because sometimes that is not easy to define.

30. *The gates through which the bonding agent attacks.* There are three gates through which the hunter of souls ventures to bind: vision, hearing, and mind or imagination. If it happens that someone passes through all three of these gates, he binds most powerfully and ties dowm most tightly. He who enters through the gate of hearing is armed with his voice and with speech, the son of the voice. He who enters through the gate of vision is armed with suitable forms, gestures, motions and figures. He who enters through the gate of the imagination, mind and reason is armed with customs and the arts. After that, the first thing that happens is the entrance, then the joining, then the bonding, and fourthly the attraction. The one who is bound encounters the bonding agent through all the senses, up to the point that a perfect bond has been made such that the former is totally immersed, and desires to be totally immersed, in the latter. And thus, a bond of mutual desire is established. Parallel to this, there are, indeed, also unpleasant bonds, which we will discuss later when we talk about natural bonds. For example, the toad attracts the weasel with a hidden power in its breathing; the cock overwhelms the lion with its voice; the mullet, by its touch, stops a ship; in his fantasy, the fanatic devours the demon; and a melancholic and unstable humour acts like a magnet on evil spirits.

In conclusion, there are thirty topics which relate to the general theme of the bonding agent, namely:

1. Types
2. Effects
3. Art
4. Rank
5. Steps
6. Multitude

7. Talent
8. Power
9. Coincidence of contraries
10. Diversity
11. Mediation
12. Partiality and concurrence of circumstances
13. Instruments
14. Opportunity
15. Differences
16. Variable powers
17. Location
18. Predisposition
19. Diversity of predispositions
20. Condition
21. Reaction
22. Distinction
23. Blindness or ignorance
24. Diligence
25. Weapons
26. Vicissitudes
27. Eyes
28. Enticements
29. Sequences
30. Gates

On what can be bound in general

1. *Types of things which can be bound.* There are four things which rotate around God, or universal nature, or the universal good, or absolute beauty. They rotate in such a way that they cannot abandon that centre, otherwise they would be annihilated, and in such a way that they can be separated from that centre only by the distance of each of their circumferences from its proper centre. These four things, I say, move in a circle around their bonding agent in such a way that they maintain the same order forever. According to the Platonists, they are mind, soul, nature and matter. Mind, in itself, is stable; soul, in itself, is mobile; nature is partly stable and partly mobile; and matter, as a whole, is both mobile and stable.

2. *The condition of that which can be bound.* Nothing is bound unless it is very suitably predisposed, for that brightness[5] is not communicated to all things in the same way.

3. *The form of that which can be bound.* Everything which is bound has an awareness in some sense, and in the nature of that awareness, one finds a certain type of knowledge and of appetite, just as a magnet attracts or repels different kinds of things. Hence, he who wishes to bind ought to focus in some way on the awareness in that which can be bound. For, indeed, a bond accompanies the awareness of a thing just like a shadow follows a body.

4. *The comparison of things which can be bound.* Let us note that humans are more open to bonding than are animals, and ignorant and stupid men are very much less suited for heroic bonds than are those who have developed an illustrious soul. In regard to natural bonds, the common person is much more susceptible than is the philosopher; as the proverb says, the wise rule over the stars. In regard to the intermediate type of bonds, it happens that the greedy person might boast of being temperate, and the lustful person of being moderate.

5. *The distinction of things which can be bound.* From what has just been said, it must be noted that the strength of one bond makes another type of bond less forceful or more mild. Thus, a German is less agitated by Venus, an Italian by drunkenness; a Spaniard is more prone to love, a Frenchman to anger.

6. *The seed or incitement of the capacity to be bound.* A thing is bound in the strongest way when part of it is in the bonding agent, or when the bonding agent controls it by one of its parts. To show this with just one example, necromancers are confident that they exercise control over entire bodies by means of the fingernails or the hair of the living, and especially by means of footprints or parts of clothing. They also evoke the spirits of the dead by means of their bones or any part of their bodies. Hence, it is not accidental that special care is taken in burying the dead and in preparing funeral pyres, and that leaving a body unburied is counted among the most grievous crimes. Also orators create good will with their art when their listeners and judges find something of themselves in it.

[5] For Bruno's use of this term, see Part Three, 'On Cupid's Bond and on Bonds in General', paragraph 1, 'The definition of a bond'.

7. *The timing of the capacity to be bound.* In different seasons and ages, one and the same thing can be bonded in various ways, and different things are not related to one and the same bond in the same way. Nor are wholes always recomposed in the same way. From this we can point out that someone who was easy going and showy as a young man becomes a more stable and prudent adult, while an old man is more suspicious and morose, and a very old man is full of blame and loathing.

8. *The diffences of things that can be bound.* Whoever wishes to bind must take note of the fact that some of the things that can be bound are affected more by nature, others more by judgement or prudence, and still others more by practice and habit. As a result, the skilful person obliges and binds the first type of things with bonds provided by natural things, the second type by reasons and proofs, by symbols and arguments, and the third type by what is at hand and is compelling.

9. *Resistance to being bound.* The more that a soul is bound to one object, the more it turns away from and rejects others. Therefore, he who wishes to limit what can be bonded to only one bond should make a special effort to make it insensitive to other activities and objects, and to turn it away from any concern for them. For, indeed, a more pleasant action excludes a less pleasant one; the soul that is intent on hearing neglects vision; he who observes more attentively becomes deaf; when we are either very happy or sad for some reason, we are little concerned with the other; when we are lazy we stop or slow down our work, that is, we become restrained, pulled away, held, bonded. As a result, when the orator breaks the bond of love by laughter or envy or other feelings, he binds by hate or contempt or indignation.

10. *The number of things that can be bound.* Thinking persons turn away from sensible things and are bound by divine things. Pleasure seekers descend through vision to the abundances of touching. Moralists are attracted by the amusement of conversation. The first are heroes, the second are natural, and the third are rational. The first are higher, the second lower, the third in between. The first are said to be worthy of the heavens, the second of life, the third of feeling. The first ascend to God, the second cling to bodies, the third move away from one extreme and approach the other.

11. *The motion of what can be bound.* All composite and variable things, and generally all things which undergo changes in their nature and disposition,

such as the soul and the spirit, are subject to various changes in their bodies and in the motions of their bodies (for although each substance is quite stable and eternal because of its simplicity, still it acquires a desire from its privation, an impulse from its desire, a motion from its impulse, and a breaking of bonds from its motion). As a consequence, no bonds are eternal. Rather, things alternate between bondage and freedom, between being bonded and escaping from a bond, or they transfer from one type of bond to another. This is a natural occurrence, and it precedes, accompanies and follows the eternal condition of all things. Thus, nature binds with its variety and motion, and art, which emulates nature, multiplies, varies, diversifies, orders and arranges bonds in a successive series. But complete stability is opposed to the nature of things, just as we are sometimes more inclined to condemn it, and yet at other times we rather desire it, for it is quite natural to desire to break from bonds, while just a little while ago we were open to being tied to them by our own voluntary and spontaneous inclinations.

12. *The indefiniteness of what can be bound.* Insofar as that which can be bound is composed of more parts, to the same degree it is less limited to specific bonds. Thus, human pleasure is less limited to only one time or individual or sex than are the pleasures of animals. All horses would have an equal chance to mate with one mare, but this is often not true of all men and one woman. This separation and indeterminateness between humans and animals is also found between a true human and a brutal human, between a more sensitive, and also more feeling, person and a more dull person. And what we have said about one type of bond must also be applied to all other genera and species of bonds.

13. *The foundation of the capacity to be bound.* The primary reason why each thing is capable of being bound is partly because there is something in it which strives to preserve itself as it presently is, and partly because it strives to be completely developed in itself according to its circumstances. In general, this is self-love. Hence, if one could extinguish self-love in an object, it would be subject to any and every type of bonding and separation. On the other hand, when self-love flourishes, all things are easily attracted to the types of bonds natural to them.

14. *The relation of things that can be bound.* Consider the friendship and the enmity among animals, their sympathy and hostility, their similarity and

diversity, and the circumstances of such things. Then arrange in an order and in an analogy all the particularities and the separate individuals in the human species, then all of the individuals and all the species of the other animals, and finally the species of all other things. You now have collected before you in a convenient order the diversity of bonds.

15. *The material diversity of things that can be bound.* Although everything that can be bound is composite in some way, still one thing can be said to be simple and another many-sided or complex, and one thing can be more simple while another has more parts. Consequently, some things are bound purely and others impurely, and some bonds are pure while others are impure. Thus some pleasures and pains are pure, some are impure, and some are mixed. For example, Epicurus taught that the pleasures of Venus are impure, because they are accompanied by pain and by an insatiable desire (by which the whole body tries to transform itself into another whole body), and this results in a sorrowful exhaustion. If there are things whose principles never fail (perhaps the stars and the great living souls or gods of the world, in whom there is no fatigue and in whom the influx and outflux of substance is always exactly the same), then they would be bound by themselves to each other in the most happy way.

Therefore, he who desires to bind in a socially effective way must take into account the diverse composition or structure of things, and must consider, evaluate and decide differently when dealing with heroes, or with ordinary people, or with those who are more like brutes.

16. *The degrees of things that can be bound.* Children are less bound by their natural feelings, because their nature is absorbed in growth and is disturbed by great changes, and all their nutrition is given over to growth and the structuring of the individual. But they clearly begin to be open to being bound in the fourteenth year, for even though at that age they are still involved in growth, their rate of growth is not as fast and as great as when they were children. And in the stable period of adulthood, men have a greater strength in their semen and, as a result, seem to be more subject to being bound. Furthermore, adolescents and young men seem to be more sexually excited for the reason that they are on fire for a long time because of the novelty of this pleasure; because the passages through which the semen passes are narrower, the wetness gushes forth with a more delightful pleasure. And as a result of the sexual itch which arises from this pressure, they are more delighted and liberated. But bonds are more difficult

in older men, whose powers are half dead, whose organs and passages are spent, and whose semen is not abundant. Precisely the same thing is found proportionally in the other emotions which have an analogy or contrast or dependence on the passion of love.

17. *The temperaments of things that can be bound.* Because of their temperament, those who are melancholy are more bound to indignation, sadness, pleasure and love, for since they are more impressionable, they also have a stronger sense of pleasure. They are also more prone to contemplation and to speculation, and in general are moved and agitated more often and more strongly by their emotions. Hence, in regard to the affairs of Venus, they regard pleasure as an end in itself rather than as a means to propagate the species. Next to them are people who are choleric, in comparison to which the sanguine are less agitated. Those who are phlegmatic are less lustful than the others, but are more greedy. Nevertheless, the fact remains that everyone has his role in obeying nature. The melancholy are bound by a greater force of imagination; the sanguine by a greater ability to emit sperm and by their hot temper; the phlegmatic by their greater abundance of fluids; and the caloric by their being more strongly and more sharply agitated and stimulated by a hot spirit.

18. *The signs of things that can be bound.* Physiognomy also has its part to play in these considerations. There are people who have slender and sinewy tibias, and who are similar to goats and to satyrs in having a wide concave nose, deep breathing and a languid face. Such people love more intensely and pursue sexual license more strongly. At the same time, they are easily appeased and do not have any emotion for a long time.

19. *The duration of things that can be bound.* With respect to bonds, old men are more stable but less suitable; young men are more unstable but more suitable; but middle aged men are bound suitably, skilfully and in a stable manner.

20. *The reaction of things that can be bound.* Mutual agreeableness gives rise to mutual bonds. Thus, there are bonds in jokes, in wit and in theatrical performances. In these ways, even those who are ugly and deformed can bind those who have feelings for them. Let us add that we have often tried to think about what it would be like to have a huge and lustful body, since the following imaginary verses were composed to be cast as a spell upon a young boy or girl:

I confess that I lack a beautiful form;
Yet God prefers me as more excellent,
As does a girl who is steady and not silly.[6]

In a proportional way, there are bonds by which those who are ugly bind because of their reputation for courage, vigour, eloquence, ingenuity and other such things, for from one type of power they can cause bonds of another type. It is not a rare occurrence that the more ugly Amazons bind in the act of love because of their reputation for their strength or their use of eloquence.

21. *The heterogeneity of things that can be bound.* Furthermore, there are species which are bound to a different species through love, hate, admiration, piety, compassion and other such feelings. For example, there are some famous cases of such bonding, like Lesbia with her sparrow, Corinna with her small dog, Cyparissus with his doe, and Arion with his dolphin. In general, the seeds of all species are attracted to other species. I will remain silent about the sympathy between a man and a lion, and I will pass over what I know about the astonishing intimacy between a boy and a snake.

22. *The changing of things that can be bound.* It is not difficult to change that which can be bound from one type of bond to its contrary, since the bonding agent is also changeable. And it makes no difference whether this occurs actually or only in thought. Even though I was once bound in thought by a teaching, the bonds of contempt and indignation may come later when that opinion has been studied in a better light. And the bonds arising from the fires of youth and beauty are relaxed and soothed in time when they do not agree with the bonds derived from customs and skill.

23. *The cause and effect of things which can be bound.* What it is that bonds to love and hate or contempt is hidden to the functions of reason. Adrastia's explanation is useless: namely, that the explanation of love, which arises from seeing a beautiful object, is a recollection by the soul of divine beauty, which was first seen as a companion to the body. If this were true, what is

6 Bruno quotes these three lines, with minor variations, from the 'Priapea', a set of approximately eighty short Latin poems, mostly from the Roman era, which were collected by an unknown editor. The poems commemorate Priapus, the mythological Greek god of plant and animal fertility, who was depicted as having a grotesque body and an enormous phallus. The 'Priapea' can be found in Emil Baehrens, ed., *Poetae latini minores* (Leipzig: Teubner, 1879) Vol. 1, 58–87. Bruno's quotation is from #38 (70).

it that suddenly changes the soul to reject an object which in no way has changed in its nature? Why are different souls captured more easily by different objects? Why does that which is beautiful to one person also turn out to be ugly to another, no less talented person? Thus, the condition of things that can be bound is unfathomable to casual and routine examination.

24. *The definition of things that can be bound.* Theocritus attributed love and the other emotions by which individuals are bound to luck or fate or some kind of an indeterminate cause. But he would have understood them more clearly if he had thought and said 'a hidden but determinate' cause instead of an 'indeterminate' one, because their origin is not apparent. The emotions, indeed, have a definite and rational structure which is either given by nature or which arises from practice and habit.

25. *The meaning of things which can be bound.* The fact that things are bound by love or hate or some other feeling was explained by the Achaeans as due to fate and not to reason or to any type of thought. As a result, they venerated both love and hate at the same altar. Some Platonists agreed with this opinion, saying that animals who cannot speak are never bound by love, because they lack reason and prudence. But they had too limited a view of the nature of thought and reason, which fills all things with the universal spirit and which shines forth in all things and proportionally in each object. But, for us, love, like all emotions, is a very practical form of knowledge. Indeed, it is a type of discourse and reasoning and argumentation by which humans are most powerfully bound, even though it is never listed among the primary types of knowledge. Therefore, he who wishes to bind believes that reason has neither a greater role nor a more important role than love in binding, although indeed the latter falls under the genus of knowledge.

26. *The flight of that which can be bound.* Sometimes that which ties itself by one type of bond flees in order to bind itself with another type of bond. As a result, he who wishes to bond should be careful to use means which effectively bind the object, that is, he should employ the bonds which already hold it. For example, a hunter, who was absorbed by his interest in and attention to wild animals, was sexually seduced with appropriate gifts by the nymph who used a horn whose sound made fleeing animals come to a standstill. Also a soldier may be bound by other feelings by enchanting him with the power of his weapons. Thus, people are bound to sex through hunting, fasting, drunkenness, gymnastic exercises and in general through

various types of concerns, leisure activity, abstinence, debauchery, etc. And what is true of this type of bond also applies to the others in their own way.

27. *The substance of that which can be bound.* There are two causes of the act of bonding, and they are the same as the two causes of the essence of that which, as such, can be bound: namely, knowledge of some kind and desire of some kind. If something has no desire at all, then it cannot be bound spiritually in any way at all. Furthermore, if something has no knowledge and desire, then it cannot bind anything either socially or through magic. I will not speak of the other types of bonds because I would not want to say anything unsuitable to those of limited vision, who are numerous.

28. *The completeness of that which can be bound.* Something is perfectly bound if it is bound in all its powers and components. Hence, he who binds should count these items carefully so that, in wishing to bind as completely as possible, he can tie up many or all of them. He should have no doubt or confusion about the different types and powers of nourishment and enticement which affect the soul and the spirit.

29. *The connection of things that can be bound.* It is not possible for a bonding agent to bind something to himself unless the former is also bound to the latter, for bonds adhere to, and are inserted into, that which is bound; the bonding agent, which may accidentally be bound to another object, must be truly bound to the object which it binds to itself. However, the bonding agent has an advantage over that which is bound, for he is master over the bonds, and because he is not affected and influenced in the same way. This notion is supported by the fact that a procurer binds but is not bound, but she who is loved is not bound by her lover unless he is bound by her in the same act of love. Furthermore, a spiritual and mysterious type of bond occasionally also occurs in which she who is loved is bound by her lover, but she neither knows nor loves him. This is the type and level of love in which Eros was brought to tears and unhappiness by Anteros. But at the social level, no one binds unless he is also bound by the same or a similar type of bond either to someone, or at least with someone, whom he desires to bind.

30. *The truth of that which can be bound.* For that which can be bound to be truly bound, a real bond is not required, that is, a bond which is found in things. An apparent bond is enough, for the imagination of what is not true can truly bind, and by means of such an imagination, that which can be

bound can be truly bound; even if there were no hell, the thought and imagination of hell without a basis in truth would still really produce a true hell, for fantasy has its own type of truth. It can truly act, and can truly and most powerfully entangle in it that which can be bound, and thus the torments of hell are as eternal as the eternity of thought and faith. As long as the soul, even when stripped of the body, retains these same characteristics, it maintains its unhappy state for ages, and perhaps even more so because of its pleasures and drinking and lack of self-control. The common philosophers did not understand this, and they most stupidly used this teaching to condemn the most ignorant of people. We will not make a big issue of this, except to say the following: when we were children and inexperienced, we were flooded with the arguments of these philosophers, just as much as the old and the experienced, themselves, had been flooded with the same arguments. Nevertheless, we forgive these elders for these views, just as much as we think that we should be forgiven, since we were just children.

On cupid's bond and on bonds in general

We have claimed in our treatise *De naturali magia*[7] that all bonds are either reduced to the bond of love, depend on the bond of love or are based on the bond of love. An examination of our thirty topics of discussion will easily show that love is the foundation of all feelings, for he who loves nothing has no reason to fear, to hope, to praise, to be proud, to dare, to condemn, to accuse, to excuse, to be humble, to be competitive, to be angered or to be affected in other ways of this sort. Hence, in this section, which we have entitled 'On Cupid's Bond', we have the opportunity to deal with a topic which is very familiar and with considerations and speculations which range very widely. This examination should not be considered to be far removed from public affairs just because it is more important and more wonderful than the field of public affairs.

1. *The definition of a bond.* According to the Pythagoreans and the Platonists, the bond of beauty is said to be a brightness, a beam of light and a certain motion, or at least its shadow and image and trace. It has spread out first into the mind, which it adorns with the order of things; second into the soul, which it brings to completion with the sequence of things;

[7] See *On Magic*, 'On the Analogy of Spirits', #2.

third into nature, which it diversifies and sustains with its seeds; and fourth into matter, which it supplies with forms. According to them this beam of light is clearest in the mind, clear in the soul, obscure in nature and most obscure in the material substrate of natural things. It is not a bodily mass, and it has no bulk. Nor can it rotate around a mass and through the whole of space, for not just large things, but also small ones, are seen to be beautiful. In the same species, large things are deformed and small things are beautiful, but the opposite also occurs, and it often happens that beauty is lost when something remains the same in quantity, and is preserved when that quantity is changed. The most beautiful baby or child is pleasing but does not bind until he is an adolescent of a certain age. Then he has some size, and this is true even if his form and figure and complexion have not changed at all. From this we conclude that social types of bonding require a degree of size on which the form and the power of the bond depends. It refers, I think, to gestures, words, clothing, habits, sense of humour, and the other signs of human feelings.

2. *The origin of a bond.* Some Platonists define a bond as arising from a certain proportionality of parts accompanied by a certain pleasantness of colouring. But to those who consider the matter more fully, it is at least as clear that it is not just composite things and things consisting of parts that bind, but that colour alone and sound alone also bind. Furthermore, nothing slips away and ages faster than beauty, and nothing changes more slowly than the form and figure which shine forth from the composition of parts. Hence, it seems that the bond of beauty must be sought elsewhere than in the figure and in the proportionality of parts. Indeed, sometimes love passes away after the flowering of the object loved, but the same beauty and figure still remain. As a result, the nature of a bond consists chiefly in a certain mutual orientation between a captor and a captive. Indeed, it sometimes happens that even though we have no grounds to complain reasonably about a girl's beauty, or in a social setting to criticize someone's conversation, speech, habits or actions in general, still they do not please us. On the other hand, something, or even many things, may displease us in someone, yet we still love that person. And, indeed, it would be rather stupid to identify colour as a bond without distinguishing between colour and the things associated with colour. For does colour in itself bind when a brighter colour is displeasing and rejected by an old man, while a duller colour binds and captures a young man? And also if, in a social setting, an

adolescent were to give a serious speech about grave matters of state, then no matter how brilliant the speaker's oratory, a man of more mature judgement would become indignant because of the speaker's arrogance. Likewise, if an old man were to give a speech full of charming, flattering and flowering words, this would invite contempt, would sometimes provoke laughter and would provide an occasion for mockery. Thus, in regard to the body, to words and to behaviour, one thing is fitting for a married woman, another for a virgin; one thing for a girl, another for a boy; one thing for a mature adult, another for an old man; one thing for a soldier, another for a Roman citizen.

3. *The indeterminateness of a bond.* I believe that it is not as difficult to make and to break bonds as it is to identify a bond in the concrete circumstances in which bonds are referred to the case at hand rather than to nature or to art. For example, a bond which originates from the body has no specific location in the body. Consider the eyes and cheeks and mouth by which a lover feels that he is bound. When these same things are attributed to another subject in the same proportions, it sometimes happens that they do not bind in a similar way, and thus the bonds of Cupid are dissolved or prevented. Why is it that sometimes, when we are consumed by love for a body which we have seen, the bonds of Cupid vanish when we become acquainted with that individual's speech and personality? And thus, you should understand bonds in the same way in a social setting.

4. *The composition of a bond.* The bond of Cupid is inferior to the bonds by which appropriate composite things bind us, and we are in no way forcefully captured by simple and absolute things. There are those who strongly reject these latter bonds. They think that God has no beauty in Himself, since his nature is simple and He does not display any level of composition. However, it is a matter of faith that God is both the author and the goal of all beauty and of every bond. Thus, because of the weakness of their minds, these thinkers have not distinguished between beauty in itself and what is beautiful to us. Likewise, at the practical level, they do not discern and distinguish between what is beautiful and reasonable to all men on the one hand, and what is a matter of custom, practice and opportunity for particular people on the other hand. As a result, they err in their attempts to bind.

5. *The number of bonds.* To put the matter generally and firmly, bonds are the form, the habits and the motions of a body, the consonance of voice and

speech, the harmony of customs and the chance meetings of sympathy by which men are bound to men, animals to animals and even animals to men. Thus, it is clear that the sight of a snake raises a mortal fear in a child, and the sight of a wolf terrifies a lamb, because of a natural prompting and not because of any previous experience or acquaintance, while the sight of a cow or a sheep causes playfulness and enjoyment. There are also various aromas by which men and spirits are bound in various ways. I have known men who were so terribly horrified by the smell of musk, which is sweet for everyone else, that they have even fallen down because their spirit was so disturbed. And I have known one person who was extraordinarily delighted by holding under his nose a bug crushed by his fingers. Thus, different things are bonded in different ways, and not only contrary but even diverse things are bonded to each other. Furthermore, at the social level, it is clear that Germans and Italians do not have the same language, or the same habits of caring for and clothing the body, or the same grace and elegance in their customs. Nevertheless, an individual Italian may diverge from his national norm and be more like a German; vice versa, a German may be more like an Italian. This causes a complication and requires great prudence in binding at the social level, especially when the bonds are cast not over a group but over an individual. Indeed, it is easier to bind many rather than only one. A hunter has a greater chance of hitting a bird with an arrow shot into a group of birds than he would have of hitting a particular bird with a more accurate aim.

6. *The gates of bonds.* The senses are the entrances through which bonds are cast. And vision is the most important of them all. The others are more suitable for different objects and powers: touch is bound by the softness of the flesh, hearing by the harmony of sound, smell by the sweetness of breath, soul by the elegance of customs and intellect by the clarity of proofs. Different bonds enter through different windows; they have different effects in different people, and they please because of different desires in different people, for, indeed, a bond does not arise equally from all things, nor have an effect equally in all things.

7. *The types of bonds.* We know that there are as many types of bonds as there are types and varieties of beauty. Also, these varieties do not seem to be smaller than the primary varieties of things, that is, the different species. Furthermore, within each species there are different individuals who are bound by different things in different ways. Thus, the hungry are bonded

to food, the thirsty to drink, he who is full of semen to Venus; one person to a sensory object and another to an intellectual object; one person to a natural object and another to an artificial one; a mathematician is bonded to abstractions and a man of action to concrete things; a hermit satisfies himself by a desire for what is absent and a member of a family by what is present. Different things are bonded by different things in every species, and the same bonds do not of themselves carry the same power when they originate from different sources. Bonds arise when music is played by a boy or an adolescent, but less so if by a girl or a man. Strength in a man is binding because of his great size, but not in a woman. A girl binds through simplicity and honesty, but if an adult has the same influence, bonds are broken and he is more and more displeasing.

8. *The measurement of bonds.* At the social level, orators, court officials and those who know how to get things done bond more effectively if they secretly conceal their skills when they act, for he who speaks with too much eloquence, or who displays a knowledge too full of trivia, will not be well received. Those who dress too rigidly and too precisely are displeasing, and so is curled hair, and eyes, gestures and motions which always follow a precise format, while he who keeps himself far removed from such things is not displeasing. Public speaking of this type is generally thought to be too affected and too florid. This is due to laziness and to a lack of talent and of good judgement, for to conceal an art while using it is no small part of the art. Thus, he who eloquently displays his knowledge at all times on every topic is not very wise, just as one who has rings and jewels on all of his fingers is not well adorned, and one who arrives loaded down with many different necklaces is not well dressed. From this we should especially realize that a bright light extinguishes a bright light, and that without darkness, light does not shine, gleam, glitter and please, for an ornament is nothing when it does not complement that which it adorns and shapes. Thus, art is not separated from nature, nor is culture foreign to simplicity.

9. *The description of a bond.* For Plato, a bond is a type of beauty or agreement of forms; for Socrates, it is the excelling charm of the soul; for Timaeus, the tyranny of the soul; for Plotinus, the private law of nature; for Theophrastus, a secret deception; for Solomon, a hidden fire and furtive waters; for Theocritus, a precious destruction; for Carneades, an agitated ruler; and for me, 'a joyful sorrow, and sorrowful joy'.[8] From what we have said in the

[8] In other writings, Bruno uses this phrase to refer to himself.

preface to this part of this treatise, these other descriptions of feelings and other types of bonds have an analogy to our notion of feelings and bonds.

10. *The distribution of bonds.* Perfect things are bonded to perfect things; noble things and nobility are bonded to noble things; and things which are imperfect and defective are bonded to things which are imperfect and defective. As a result, it was said above that part of what is in that which is to be bound must be present in the bonding agent. A completely chaste girl, in whom there are no seeds of excitement, is not bound to sensory pleasure by any star or by any artifice if she has not been touched or embraced, that is, (I say) she has not submitted herself to the hand of a bonding agent, and his hand has not reached out to her. I will say nothing about an immature girl, for in all actions there must be some seed, but not all seeds are fruitful everywhere. And whose attempt to entice someone who is ill, or old, or frigid or castrated would not be frustrated (the opposite would apply to those who would not make the attempt)? In regard to social bonds, a proportional judgement is quite easy to make.

11. *The degrees of bonds.* Things in the universe are so ordered that they constitute one definite co-ordination in which there can occur a transition from all things to all things in one continuous flow. Some of these things are immediately related to others, for example, the natural propagation of individuals of the same species, and in these cases the bonds are blood related, familiar and easy. Other things are interrelated through various intermediaries, and all of these intermediaries must be crossed over and penetrated so that bonds are stretched across from the bonding agent to that which is bound. Thus, by their generosity to things and by their good-will in sharing with these intermediaries, spirits influence inferior, and even the lowest, things and bind them to themselves. On the other hand, lower things are raised up with a certain reverence through a natural or rational sequence so that, through the free consent of higher things, they can bind to themselves superior things located far above them. And just as there are various species of things and differences between them, they also have various times, places, intermediaries, pathways, instruments and functions. It is very easy to see this and to understand it for all types of bonds and things that can be bound.

12. *The size of a bond.* In all things there is a divine force, that is, love, the father himself, the source, the Amphitrite of bonds. Thus, Orpheus and

Mercury were not wrong when they called this the great demon, for this bond is indeed the entire substance, constitution, and (if I may say so) the hypostasis of things. We come to know this greatest and most important bond when we turn our eyes to the order of the universe. By this bond, higher things take care of lower ones, lower things are turned toward higher ones, equal things associate with each other and lastly, the perfection of the universe is revealed in the knowledge of its form.

13. *The principal effect of a bond.* If there were only one love, and thus only one bond, all things would be one. But there are many different characteristics in different things. Hence, the same thing binds different things in different ways. As a result, Cupid is said to be both above and below, both the newest and the oldest, both blind and most observant. Cupid made all things in such a way that, for the preservation of their species, they remain firm in their powers or in themselves and are not separated from themselves. But then, in regard to the changes which occur in individual things, he arranged it so that they would be separated from themselves in a certain sense when the lover eagerly desires to be completely transported into the loved one; and also that they would be unrestrained, opened up and thrown wide open when the lover desires to embrace and to devour the loved one completely. Thus it happens that the bond by which things wish to be where they are and not to lose what they have also causes them to wish to be everywhere and to have what they do not possess. This is due to a sense of complacency with what is possessed, to a desire and an appetite for what is absent but possessable, and to a love for all things. A particular and finite good and truth is not sufficient for an individual appetite and intellect, which have as their objects what is universally good and universally true. From this it follows that a finite potency in some definite material body simultaneously experiences the effects both of being drawn together and of being pulled apart, dispersed and scattered by the same bond. This general characteristic of a bond is to be found in each individual type of bond.

14. *The quality of a bond.* In itself, a bond is neither beautiful nor good. Rather, it is the means by which things as a whole, and each individual thing, pursue what is beautiful and good. It connects that which receives with that which is received, that which gives with that which is given, that which can be bound with a bonding agent, that which is desired with the one who desires. Indeed, that which desires the beautiful and the good

lacks these qualities insofar as it desires them. Thus, to that degree it is neither beautiful nor good. Hence, one of the Peripatetics was wrong in his statement about matter when he concluded that matter is ugly and evil, because the desire for the good and beautiful is itself evidence that matter lacks these properties. Aristotle said more carefully that matter is not 'ugly' or 'evil' as such.[9] Rather, the actual truth is that that which, like matter, tends and moves equally towards goodness and evil, ugliness and beauty, is in itself neither ugly nor beautiful, neither evil nor good. If matter were evil, it would be contrary to its nature to desire the good; the same would be true if it were naturally ugly. And if it were evil by analogy, then it would also analogously possess a contrary which does not desire, but, rather, excludes and rejects, the other contary. The more profound philosophers understand this as we have declared elsewhere. That is, matter itself, in its bosom, is the beginning of all forms, such that all things originate and are produced from it; it is not a pure negation, as if all things originated from the outside as foreigners; indeed, outside of the bosom of matter there are no forms; rather, all forms are both latent within it and are derived from it. Consequently for anyone who considers bonds at the social level and in their full meaning, it should be clear that in every material thing or part of matter, in every individual or particular thing, all seeds are contained within and lie hidden there, and, as a result, the inclinations of all bonds can be actuated by a skilful effort. In one of our 'Thirty Small Signs',[10] we have explained in general how such an inclination and its transformation take place.

15. *The generality or universality of a bond.* From what has just been said, it follows that the love by which we love, and the tendency by which all things desire, are intermediaries between good and evil, between the ugly and the beautiful (not themselves being ugly or beautiful). And so they are good and beautiful because of a sort of sharing and participation, for the bond of love has a nature which is both active and passive. And by this, things act, or are acted upon, or both, as they desire to be ordered, joined, united and completed, insofar as it is within the nature of each thing to be occupied with order, joining, union and completion. Without this bond there is nothing, just as without nature there is nothing. Because of this, therefore, love is not a sign of imperfection when it is considered in matter and in the

9 See *Physics*, I, 9 (192.a.24).

10 See Part I, 'On Bonding Agents in General', Article 12, 'No one particular thing can bind everything'.

chaos before things were produced. For indeed, anything which is considered in the chaos and in brute matter, and is also said to be love, is simultaneously said to be a perfection. And whatever is said to be imperfect, disordered and not to be, is understood not to be love. Thus, it is established that love is everywhere a perfection, and this bond of love gives witness everywhere to perfection. When an imperfect thing desires to be perfected, this, indeed, takes place in something which is imperfect, but not because it is imperfect. Rather, this happens because of a participation in a perfection and in a divine light and in an object having a more eminent nature, which it desires more strongly inasmuch as the object is more vivacious. That which is more perfect burns with greater love for the highest good than does that which is imperfect. Therefore, that principle is most perfect which wishes to become all things, and which is not oriented to any particular form but to a universal form and universal perfection. And this is universal matter, without which there is no form, in whose power, desire and disposition all forms are located, and which receives all forms in the development of its parts, even though it cannot receive two forms at the same time. Hence, matter is in a sense divine, just as a form, which is either a form of matter or nothing, is also in a sense divine. There is nothing outside of matter or without matter, otherwise the power to make and the power to be made would be one and the same thing, and would be grounded in one undivided principle, because the power to make anything and the power of anything to be made would be either present or absent together. There is only one potency taken absolutely and in itself (whatever it may be in particulars, in composites and when taken accidentally, a question which dominated the thinking and the minds of the Peripatetics and their monkish followers). I have said this in many places in my *De infinito et universo* and more precisely in my *De principio et uno*[11], where I conclude that it is not a foolish opinion which was defended by David of Dinant and by Avicebron in his *Fons vitae*, who cited the Arabs who also ventured to assert that God is matter.

16. *The comparison of bonds*. The most important of all bonds is the bond of Venus and of love in general, and that which is primarily and most powerfully the opposite of love's unity and evenness is the bond of hate. Indeed, to the degree that we love one of two opposites and contraries of any type,

[11] For the former (*De l'infinito universo e mondi*), see G. Bruno, *Dialoghi italiani*, ed. by G. Gentile and G. Aquilecchia (Florence: Sansoni, 1957) 382–9; for the latter (*De la causa, principio, e uno*) see *Ibid.*, pp. 182, 262, 315. (above, pp. 7, 55, 86)

then to that same degree we hate and reject the other. These two feelings, or rather, in the last analysis, this one feeling of love (whose substance includes hate) dominates all things, is lord over all things, and elevates, arranges, rules and moderates all things. This bond dissolves all the other bonds. For example, female animals who are restrained by the bond of Venus do not get along well with other females, and males do not tolerate rival male suitors. They neglect food and drink and even life itself, not giving up even when conquered. Rather, the more they are worn out, the more they press on, fearing neither storms nor the cold. Because of this argument, Aristippus decided that the highest good is bodily, and especially sexual, pleasure, but he held before his eyes a rather animalistic view of man as a result of his own conclusion. But still, it is true that the more skilful and clever bonding agent, who uses things which the one to be bound or tied loves and hates, expands his pathway to the bonds of the other feelings. For indeed, love is the bond of bonds.

17. *The time and place of bonds.* Even though the best seed is sown, the generation of new things does not occur always and everywhere. Likewise, bonds are not effective always and everywhere in capturing an object, but only at the proper time and with the appropriate disposition of the object.

18. *The distinction of bonds.* There are no purely natural or purely voluntary bonds (in the sense in which people commonly distinguish between the natural and the voluntary). The will acts with the participation of the intellect, while the intellect is not limited by the will but acts everywhere, except where nothing exists. We have proven this in other places, and thus it would be useless to discuss the matter further here. According to our understanding, there are three different types of bonds: the natural, the rational and the voluntary (even though all things are based on one natural foundation). Consequently, to some degree we cannot set boundaries between one type of bond and another. Thus, the laws of prudence do not prohibit love, but love beyond reason. And the deceivers of the foolish prescribe without reason limits to reason, and condemn the laws of nature. And the most corrupt say that nature is corrupt, because humans are not raised above nature like heroes but are degraded like beasts as against nature and are beneath all dignity.

19. *The development and stages of a bond.* According to the Platonists, the construction of the bond of Cupid occurs as follows. First, some type of

beauty or goodness, or some such thing, is brought into the external senses. Second, it is taken on to the centre of the senses, that is, to the common sense; then, third, into the imagination; and fourth, into the memory. Then the soul, by its own power, desires first that it be moved, redirected and captured; second, once redirected and captured, it is enlightened by a ray of the beautiful or the good or the true; third, once enlightened and illuminated, it is inflamed by sensory desire; fourth, once inflamed, it desires to be united to the thing loved; fifth, once united, it is absorbed and incorporated; sixth, once incorporated, it then loses its previous form and in a sense abandons itself and takes on an alien quality; seventh, it, itself, is transformed by the qualities of the object through which it has moved and has thus been affected. The Platonists call the responses to the initial motions Cupid's preparation; the redirecting, Cupid's birth; the illumination, Cupid's nourishment; the inflaming, Cupid's growth; the union, Cupid's attack; the incorporation, Cupid's domination; and the transformation, Cupid's victory or completion.

20. *The foundation of the stages of bonds.* You can now see how this scale is based on its individual stages. Cupid's birth issues first from the body's nourishment, sensitivity and sexual expression, and second from the soul or spirit because of its charm, or playfulness, or contemplation, which is worthy of a better name, in which beauty is joined with pleasantness. Cupid's food, which prevents the newborn from expiring, is the knowledge of what is beautiful. Cupid's growth is due to a lingering reflection on the knowledge of what is beautiful. Cupid's attack consists in the fact that the soul slides and spreads from one part to all parts of the beloved so that it can inflame the whole. Cupid's domination is grounded in the action by which the soul of the lover, having abandoned his own body, lives and acts in the other. Cupid's transformation occurs when the lover, having died to himself, lives another life in such a way that he lives there as in his own house rather than in someone else's house. Thus, it is said that Jupiter was transformed into a bull, Apollo into a shepherd, Saturn into a horse and the other gods into other forms. Likewise, the soul is transformed by the motion or disturbance of its feelings from one form and type of bond to another.

21. *The condition of bonds.* There are some external things which bind, for example, gifts, acts of deference, honours and favours. But these truly bind when they are not given in such a way as to earn a favour in return. And

thus, bonds based on commercial transactions are ignoble and merely utilitarian, and are held in low esteem.

22. *The appropriateness of bonds.* The bonds which are most powerful and appropriate are those which occur through close contact with a contrary in a way which can be better explained here by examples rather than by a definition or by a name (which is unfamiliar.) Thus, consider the case in which a humble and honourable person binds a proud soul. The proud person loves those whom he perceives to praise him, and the greater the praise, the greater the love. And, in fact, the praise of a great man is more significant than that of unimportant persons, whose praise we even sometimes reject. So the one who binds carefully observes the ways in which the proud man is praised. Again, consider soldiers who wish to be known primarily for their physical strength and courage, and, as a result, are little concerned if they are not noted primarily for their wisdom and influence over things. Again, consider philosophers who glory in their knowledge of things, but are little concerned if they are not praised for their heartfelt courage. The same judgement applies to the casting of other bonds.

23. *The gratitude of bonds.* Bonds create a desire for some sort of gratitude. To give an example from one type of bond, quarrels arise between lovers when it is taken for granted that each has an obligation to the other. The lover thinks that the beloved is obliged to turn over to him her stolen soul where he, who has died in his own body, lives in another body. If the lover is less flattering to his beloved, she complains that he cares less for her. The lover complains to the beloved if …

(Bruno's text ends abruptly here in mid-sentence.)

Index

Cambridge texts in the history of philosophy

Titles published in the series thus far

Arnauld and Nicole *Logic or the Art of Thinking* (edited by Jill Vance Buroker)
Boyle *A Free Enquiry into the Vulgarly Received Notion of Nature* (edited by Edward B. Davis and Michael Hunter)
Bruno *Cause, Principle and Unity* and *Essays on Magic* (edited by Richard Blackwell and Robert de Lucca with an introduction by Alfonso Ingegno)
Clarke *A Demonstration of the Being and Attributes of God and Other Writings* (edited by Ezio Vailati)
Conway *The Principles of the Most Ancient and Modern Philosophy* (edited by Allison P. Coudert and Taylor Corse)
Cudworth *A Treatise Concerning Eternal and Immutable Morality* with *A Treatise of Freewill* (edited by Sarah Hutton)
Descartes *Meditations on First Philosophy* with selections from the *Objections and Replies* (edited with an introduction by John Cottingham)
Descartes *The World and Other Writings* (edited by Stephen Gaukroger)
Kant *Critique of Practical Reason* (edited by Mary Gregor with an introduction by Andrews Reath)
Kant *Groundwork of the Metaphysics of Morals* (edited by Mary Gregor with an introduction by Christine M. Korsgaard
Kant *The Metaphysics of Morals* (edited by Mary Gregor with an introduction by Roger Sullivan)
Kant *Prolegomena to any Future Metaphysics* (edited by Gary Hatfield)
Kant *Religion within the Boundaries of Mere Reason and Other Writings* (edited by Allen Wood and George di Giovanni with an introduction by Robert Merrihew Adams)
La Mettrie *Machine Man and Other Writings* (edited by Ann Thomson)
Leibniz *New Essays on Human Understanding* (edited by Peter Remnant and Jonathan Bennett)
Malebranche *Dialogues on Metaphysics and on Religion* (edited by Nicholas Jolley and David Scott)
Malebranche *The Search after Truth* (edited by Thomas M. Lennon and Paul J. Olscamp)
Mendelssohn *Philosophical Writings* (edited by Daniel O. Dahlstrom)
Nietzsche *Daybreak* (edited by Maudemarie Clark and Brian Leiter, translated by R.J. Hollingdale)
Nietzsche *Human, All Too Human* (translated by R.J. Hollingdale with an introduction by Richard Schacht)
Nietzsche *Untimely Meditations* (edited by Daniel Breazeale, translated by R.J. Hollingdale)
Schleiermacher *Hermeneutics and Criticism* (edited by Andrew Bowie)
Schleiermacher *On Religion: Speeches to its Cultured Despisers* (edited by Richard Crouter)

Lightning Source UK Ltd.
Milton Keynes UK
UKHW042258221119
354094UK00001B/25/P

9 780521 596589